Springer Tracts in Civil Engineering

Series Editors

Giovanni Solari, Wind Engineering and Structural Dynamics Research Group, University of Genoa, Genova, Italy

Sheng-Hong Chen, School of Water Resources and Hydropower Engineering, Wuhan University, Wuhan, China

Marco di Prisco, Politecnico di Milano, Milano, Italy

Ioannis Vayas, Institute of Steel Structures, National Technical University of Athens, Athens, Greece

Springer Tracts in Civil Engineering (STCE) publishes the latest developments in Civil Engineering—quickly, informally and in top quality. The series scope includes monographs, professional books, graduate textbooks and edited volumes, as well as outstanding Ph.D. theses. Its goal is to cover all the main branches of civil engineering, both theoretical and applied, including:

- Construction and Structural Mechanics
- Building Materials
- Concrete, Steel and Timber Structures
- Geotechnical Engineering
- Earthquake Engineering
- Coastal Engineering
- Hydraulics, Hydrology and Water Resources Engineering
- Environmental Engineering and Sustainability
- Structural Health and Monitoring
- Surveying and Geographical Information Systems
- Heating, Ventilation and Air Conditioning (HVAC)
- Transportation and Traffic
- Risk Analysis
- Safety and Security

Indexed by Scopus

To submit a proposal or request further information, please contact: Pierpaolo Riva at Pierpaolo.Riva@springer.com, or Li Shen at Li.Shen@springer.com

More information about this series at http://www.springer.com/series/15088

Giandomenico Toniolo

Introduction to Frame Analysis

First and Second Order Theories

 Springer

Giandomenico Toniolo
Politecnico di Milano
Milan, Italy

The Work is an up-to-date translation into English language of the Italian textbook "Tecnica delle Costruzioni - Calcolo strutturale: i Telai" published by Zanichelli.

ISSN 2366-259X   ISSN 2366-2603   (electronic)
Springer Tracts in Civil Engineering
ISBN 978-3-030-14666-5   ISBN 978-3-030-14664-1   (eBook)
https://doi.org/10.1007/978-3-030-14664-1

Library of Congress Control Number: 2019933202

This Springer imprint is published by the registered company Springer Nature Switzerland AG
The registered company address is: Gewerbestrasse 11, 6330 Cham, Switzerland

Preface

The present work derives from the university textbook originally drafted within the cultural tradition of the Civil Engineering School of the Politecnico di Milano and devoted to the course of Structural Analysis and Design. Form and extension of the topics are mainly dictated by the didactic necessities as developed along the teaching weeks available in the academic year.

First of all, the completeness proper to treatise has been renounced, collecting in the text the topics considered of fundamental formative value and omitting many further developments and variants for which proper references are reserved.

The presented topics have been developed examining the methodological aspects more than the factual knowledge and addressing more on the check of the assumptions, the rigour of the analysis and the consequent reliability of results.

Addressing to courses organized with regular cycles of exercises devoted to the numerical applications of the arguments presented in theoretical form during the lessons, every chapter develops an organic argument that is at the end illustrated by numerical examples in the last section.

All together, the chapters follow an overall plan that develops the topics of the structural analysis, starting from the basic criteria of the safety verifications and following on the stress calculations for the more common frame structures. Along with this route, the principal methods of the elastic analysis of frames are presented, improving more and more the assumptions up to the overcoming of the linear algorithms and to the adoption of the nonlinear analysis necessary to deal with some important "second-order" problems of the structural calculation such as the instability verifications.

The course of Structural Analysis and Design is positioned as logical development after the course of Structural Mechanics. From this discipline, the fundamental models of structural behaviour are taken, dressing them with the actual "thicknesses" due to the real construction materials. The peculiar properties of these materials and their complex construction arrangement give prominence to the reliability problems of the model: not only one solution follows, but a range of possible solutions characterized by different degrees of reliability and anyway influenced by the randomness of the input data.

Structural Analysis and Design is still limited to verification problems referred to elementary constructions for which the deduction of the model is simple. The wider problem of the design choices and of the calculation of complex construction assemblies is left to the specialized courses of the following years.

Milan, Italy   Giandomenico Toniolo

Contents

Symbols

The attempt has been to adapt the notations in this textbook to the ones more commonly used internationally in the specific disciplinary sector. A significant step forward towards the unification of notation has been done within the standardization activity carried out by associations such as C.E.B. (now fib) and C.E.C.M. The English language gives the undisputed reference, overcoming the national ones (y for yield, s for steel, etc.) and even the noblest international languages such as French (c for concrete, instead of b of béton, etc.).

However, not everything is unified and there is room for the personal preferences of different authors. Finally, interferences are not completely solved with related disciplines such as computer-oriented structural analysis.

Lists of principal meaning of symbols are reported below. The mathematical ones are omitted, taken as granted, as well as the occasional ones that continuously occur in the text and that will rely on specific preventive definitions.

Due to the high number of quantities to be treated, it is not possible to avoid repetitions and promiscuity of symbols. The context will clarify misunderstandings, and starting from the following tables, notations are divided into three different domains of application: the general one of safety criteria and actions definition for the semi-probabilistic method; the one of structural design for the analysis of frames and plates; and the one relative to the construction materials and the design of relative elements.

Despite the size of tables, the following normalized codification of symbols covers a very limited area with respect to the extent of the subject.

Note: In the following text, "1" is used for the digit one, "I" is used for the capital Roman letter, and "l" is used for the small Roman letter.

Capital Roman Letters

	Actions and safety	Structural analysis	Member design
A	Accidental action	Cross-sectional area	Cross-sectional area
B	/	/	/
C	/	/	Resultant of compr.
D	/	Diameter	Diameter
E	Effect of action	Long. elast. modulus	Long. elast. modulus
F	Action on structure	Concentrated couple	/
G	Permanent action	Tang. elast. modulus	Centre of gravity
H	/	Horizontal force	/
I	/	Second moment of area	Second moment of area
J	/	Torsional inertia	Torsional inertia
K	/	Section stiffness	Section stiffness
L	/	Total length	/
M	/	Bending moment	Bending moment
N	/	Axial force	Axial force
O	/	/	Pole, centre, origin
P	Prestressing	Concentrated load	Prestressing
Q	Variable action	Force or resultant	Longit. shear force
R	Resistance	Reaction or resultant	/
S	Internal force	First moment of area	First moment of area
T	Stress	Tors. mom. or temperature	Torsional moment
U	/	/	/
V	/	Shear force	Shear force
W	Weight of masses	Section modulus	Section modulus
X	/	Axis or unknown quantity	/
Y	/	Axis or unknown quantity	/
Z	/	Axis or unknown quantity	Resultant of tension

Small Roman Letters

	Actions and safety	Structural analysis	Member design
a	Random variab. action	Greater side dimension	/
b	/	Smaller side dimension	Cross-sectional width
c	Numerical coefficient	Numerical coefficient	Concrete cover
d	/	Flexibility	Effective depth
e	/	Eccentricity	Eccentricity
f	Probability function	Function	Material strength
g	Gravity acceleration	Function	Material density

(continued)

(continued)

	Actions and safety	Structural analysis	Member design
h	/	Height	Depth of section
i	/	Radius of gyration	/
j	/	/	Age in days
k	Probability coefficient	Stiffness	Coefficient
l	/	Length	Length, distance
m	/	Moment	/
n	Number of tests	/	/
o	(Not used)	(Not used)	(Not used)
p	Probability	Distributed load	/
q	Probability (1−p)	Variable distributed load	Unity longit. shear
r	Random var. resistance	Force (or radius)	Relaxation function
s	Standard deviation	/	Spacing
t	/	Time	Thickness
u	/	Translation along x	Perimeter
v	/	Translation along y	Creep function
w	/	Translation along z	Crack opening
x	Generic random variab.	Co-ordinate	Neutral axis depth
y	/	Co-ordinate	Distance
z	/	Co-ordinate	Internal lever arm

Small Greek Letters

	Actions and safety	Structural analysis	Member design
α	/	/	Angle (or coeff.)
β	/	Buckling coefficient	C/bxf_c ratio
γ	Partial safety factor	Shear strain	Partial safety factor
δ	/	Translation	d/h ratio
ε	/	Strain	Strain
θ	/	Angle	Angle
ι	(Not used)	(Not used)	(Not used)
κ	/	Coefficient	Ratio (or coeff.)
λ	/	Slenderness ratio	Slenderness ratio
μ	/	Friction coefficient	Specific bend. mom.
ν	/	Poisson's ratio	Specific axial force
ξ	/	Co-ord. or translation	Ratio x/h
η	/	Co-ord. or translation	Ratio y/h
ζ	/	Co-ord. or translation	Ratio z/h
o	(Not used)	(Not used)	(Not used)
π	/	3.1415927…	/

(continued)

(continued)

	Actions and safety	Structural analysis	Member design
ρ	/	Generic stress	Relaxation coeff.
σ	/	Normal stress	Normal stress
τ	/	Shear stress	Shear stress
υ	/	/	Specific shear force
φ	/	Rotation	Creep coeff.
χ	/	Shear factor	Curvature (1/r)
ψ	Combination factor	Rotation	Angle
ω	/	Instability coeff.	Instability coeff.
ϕ	/	/	Rebar diameter

Subscripts

	Actions and safety	Structural analysis	Member design
a	Acting	/	/
b	/	/	Bolt or bond
c	/	Critic, collapse	Concrete
d	Design	/	Design
e	/	/	Elast., at elastic limit
f	Actions	/	/
g	Permanent actions	/	/
h	/	Horizontal	/
i	/	ith	/
j	/	jth	At day j
k	Characteristic	/	Characteristic
l	/	/	Longitudinal
m	Material	/	Mean
n	/	Normal	/
o	/	At the origin, reference	/
p	Prestressing	/	Prestressing
q	Variable actions	/	/
r	Resistant	/	Rupture
s	/	/	Steel
t	Time	Tangent	In tension
u	/	/	Ultimate of rupture
v	/	Vertical	Viscous
w	/	/	Web
x	/	Along or around x	/
y	/	Along or around y	Yield
z	/	Along or around z	/
ε	Geometric	/	/
θ	Thermal	/	Thermal

Frequently Used Symbols

Reinforced Concrete

R_c	Concrete cubic compressive strength
f_c	Concrete cylinder compressive strength
f_{ct}	Concrete tensile strength
f_{ctf}	Concrete flexural strength
f_b	Bond strength
ε_{cs}	Concrete shrinkage
ρ_s	Geometrical reinforcement ratio (or percentage)
ψ_s	Elastic reinforcement ratio (or percentage)
ω_s	Mechanical reinforcement ratio (or percentage)

Steel

f_t	Steel tensile strength
f_y	Steel yield strength
f_{pt}	Tensile strength of prestressing steel
$f_{p0.1}$	Stress at 0.1% residual elongation (proof stress)
$f_{p(1)}$	Stress at 1% elongation under loading (proof stress)
f_{py}	Yield stress of prestressing steel
ε_t	Steel failure strain
ε_u	Ultimate strain (under maximum loading)
ε_{pt}	Ultimate strain of prestressing steel

Others

l_o	Buckling length ($=\beta l$)
$\bar{\sigma}, \bar{\tau}$	Allowable stresses
γ_C	Partial safety factor for concrete
γ_S	Partial safety factor for steel
γ_F	Partial safety factor for actions
γ_G	Partial safety factor for permanent actions
γ_Q	Partial safety factor for variable actions

Chapter 1
Criteria of Structural Safety

Abstract This chapter presents the basic criteria for safety verification, starting from the probabilistic definitions of the involved quantities, such as the action and the resistance, and proceeding with the presentation of the verification modes applicable to the frame structures, up to the detailed description of the semi-probabilistic limit states' method that is universally adopted for the structural design.

1.1 Probabilistic Interpretation of Safety

All constructions shall be designed and executed so to resist with adequate safety the actions to which they are expected to be submitted, fulfilling also the conditions necessary for their ordinary service. Furthermore, this resistance shall have an adequate duration so that the *adequate safety* and the *service conditions* may last for all the expected life of the construction.

The first requirement versus collapse refers to the *public safety*; the second one of the service conditions refers to the functional use of the construction. Both are correlated to the duration properties of the related mechanical, physical and chemical properties, also with reference to those of the actions.

Indicating with A the *action* of the external loads and with R the limit value of the *resistance*, for all the service life of the construction should be (Fig. 1.1a):

$$A < R$$

Fig. 1.1 Representation of safety verification

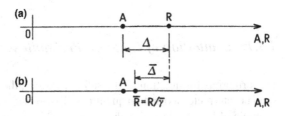

© Springer Nature Switzerland AG 2019
G. Toniolo, *Introduction to Frame Analysis*, Springer Tracts
in Civil Engineering, https://doi.org/10.1007/978-3-030-14664-1_1

But an exact value is not definable both for action and resistance. Deduced from investigations, tests, measurements and interpretations, these quantities can be defined only in an approximate way, with a certain degree of uncertainty. Therefore, the safety verification shall leave between action and resistance an opportune *safety margin* (Fig. 1.1b):

$$\Delta = R - A > \overline{\Delta}$$

as a compensation to the quoted evaluation uncertainties,

Assuming for this margin a minimum value as a ratio of the resistance level R

$$\overline{\Delta} = \overline{\delta} R$$

the verification can be written as

$$A < R(1 - \overline{\delta})$$

that is

$$\gamma = \frac{R}{A} > \overline{\gamma}$$

with $\overline{\gamma} = 1/(1 - \overline{\delta})$ minimum value of the *safety factor*. In particular, one may define

$$\overline{R} = \frac{R}{\overline{\gamma}}$$

as *admissible value* of the resistance and turn the verification in term of

$$A < \overline{R}$$

Apart from the problems met for resuming in one simple inequality the results of the structural analysis, problems that will be dealt with further on, remain in this verification criterion the limits coming from the absence of a precise quantification of safety and from the impossibility of differentiating the uncertainties of the different influencing factors.

1.1.1 Evaluation of Collapse Probability

For a rigorous formulation of the safety measure, the action A and the resistance R shall be interpreted as random quantities. To this end, one can consider the elementary example of Fig. 1.2 in which the "structure" is represented by a simple prism used

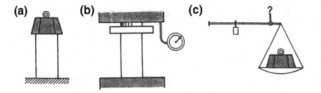

Fig. 1.2 Elementary example of the structural problem

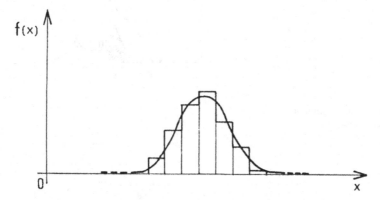

Fig. 1.3 Frequency density histogram and curve

to sustain a load (vertical force). The arrangement of Fig. 1.2a is "the structure in service" of which many samples will be produced.

For the safety verification at one side, the experimental measure of the resistance is made (e.g. by the press of Fig. 1.2b); at the other side, the measure of the action is performed (e.g. of the weight of Fig. 1.2c). Repeated for several samples, the measures x show a dispersion of results that has to be statistically interpreted. So, a *histogram of frequency density* is drawn (Fig. 1.3), dividing into parts the interval involved by the x and marking steps with a height proportional to the number of the values included in the related part. In order to allow to perform analytical elaborations, a mathematical model of the histogram is deduced represented by a continuous function: the *frequency density* curve.

In general, this model has some approximations, such as those to give small but not null densities for x values far from the interval of possible variation of x or even without physical meaning like the negative values of the resistance. Apart from these approximations, the elementary area $f(x)dx$ covered by the curve gives the frequency of the values included between x and $x + dx$ (Fig. 1.4), while the *distribution function*, defined by

$$F(x) = \int_{-\infty}^{x} f(x)dx$$

Fig. 1.4 Distribution function F and probability density f

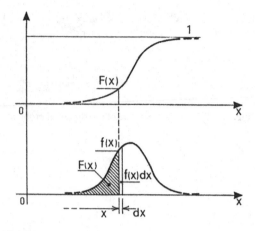

Fig. 1.5 Representation of the exact method

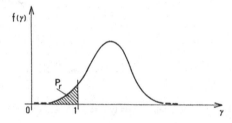

represents the frequency of values smaller than x.

If the number of measures is sufficiently high and the procedure of execution and service are the same as for the experimented samples, the same curves of Fig. 1.4 can be taken, respectively, as *probability density* and its *distribution function* related to the non-experimented structures. With these curves, a calculation for the probabilistic safety measure can be performed, that is the evaluation of the *collapse probability* as the one related to the amount of the values of the random quantity action a larger than the those of the corresponding random quantity resistance r.

Evaluation Methods

Following the *exact method*, in order to evaluate the safety, the statistical representation of the ratio $\gamma = r/a$ shall be set up, assuming for r and a the values of resistance and action of the single pairs of tests (Fig. 1.5). The collapse probability

$$P_r = P\{\gamma < 1\} = \int_{-\infty}^{1} f(\gamma)d\gamma = F_\gamma(1)$$

is represented in this case by the dashed portion of the area covered by the probability density curve of the random quantity γ.

Fig. 1.6 Representation of
the method of extreme
functionals

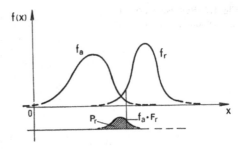

Following the method of the *extreme functionals*, the mutual independence
of the two random quantities r and a is assumed. The respective curves
$f_r = f_r(r)$ and $f_a = f_a(a)$ of frequency density are then drawn (Fig. 1.6), from which,
for any value of the abscissa x, the probability that the resistance is lower is deduced:

$$P\{r < x\} = \int_{-\infty}^{x} f_r(\xi)d\xi = F_r(x)$$

and then, assigning to the same abscissa the weight correspondent to the elementary
probability $f_a(x)dx$ of the action, one obtains the collapse probability

$$P_r = P\{r < a\} = \int_{-\infty}^{+\infty} F_r(x)f_a(x)dx$$

Once computed for the case under examination, the collapse probability is compared with the maximum value allowed on the basis of the level L of the chosen
risk

$$P_r < \overline{P_r}$$

with $\overline{P} = \overline{P_r}(L)$.

The third method here presented is called of the *extreme values* and assumes again
the mutual independence of the two random quantities r and a. From the related
frequency densities $f_r = f_r(r)$ and $f_a = f_a(a)$, two separate "characteristic values" are
provided, one R_k for the resistance, one A_k for the action. These values are defined
so to correspond to a selected probability limit p of being exceeded, respectively, in
the lower or upper direction by the corresponding random quantities (Fig. 1.7).

The verification is then set as

$$A_k < R_k$$

without evaluating the actual specific collapse probability, but implicitly imposing it
to be lower than the value

Fig. 1.7 Representation of
the method of extreme values

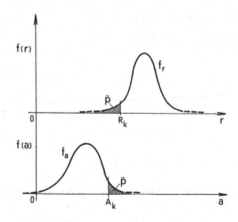

$$\overline{p_r} = \overline{p_r}(\overline{p})$$

that correspond to the limit $A_k = R_k$.

The discourse developed up to here with reference to the elementary structure of Fig. 1.1 cannot be directly extended to the real structures. Actually, these latter are usually produced in one only sample and are not experimented in situ up to collapse, but a posteriori with local indirect and not destructive methods for the indicative ends of the acceptance final test of the construction.

To obtain the resistance and action data, one performs the tests on the elementary materials at one side and adopt the loads from similar situations at the other side. The data obtained in this way go through a complex way of interpretations and calculations before arriving to the definition of a resistance capacity and an acting load that should be comparable between themselves in the verification disequality.

1.1.2 Structural Model for Design Analysis

In order to present the problem, a frame structure as that shown in Fig. 1.8 is referred to. It is constituted by an assemblage of linear elements (beams and column). For its analysis, there are three possible safety verification modes. These modes are differentiated on the basis of the different location of the comparison between resistance and action. In general, except for particular types of structures and materials, these three modes are not equivalent.

In Fig. 1.9, the synoptic scheme of the modes of concern is shown. With F_a, the loads acting on the structure are indicated (forces, thermal variations, dynamic actions, etc.). With S_a, the effects of the loads in terms of internal forces transmitted through the sections along the elements are indicated (axial and shear forces, bending

Fig. 1.8 Model of a frame structure

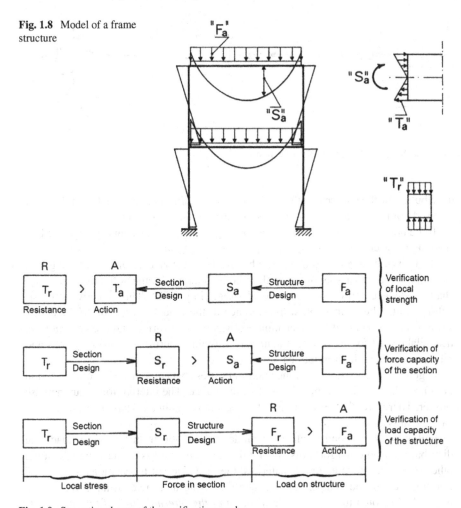

Fig. 1.9 Synoptic scheme of the verification modes

and torsional moments). With T_a, the effects of the loads in terms of local point stresses are indicated (normal and shear stresses).

At the other side, T_r indicates the resistant stress, that is the limit value deduced from the tests on the materials. S_r indicates the resistant internal force that is the limit state of the section that corresponds to its ultimate capacity. F_r indicates the resistant load that is the one that takes the structure or a part of it to the collapse limit.

Verification Modes

The first mode consists of the *verification of the local stress* on the material. One has to analyse the structure starting from the acting loads, defining the distribution of the internal forces along the elements. Subsequently for any section, one calculates the

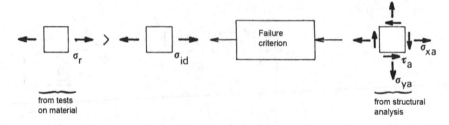

Fig. 1.10 Scheme of local stress verification

distribution of the stresses on the basis of the internal force. And finally on the most stressed points, the comparison versus the resistant stress is made.

To be noted that the local stress state is in general a tensor quantity expressed by several scalar components (σ_x, σ_y, σ_z, τ_{xy}, τ_{yz}, τ_{zx}). On the contrary, the resistant stress derives from a uniaxial test and by consequence is a scalar quantity (e.g. σ_r). To make the comparison, an equivalent homogeneous (scalar) effective stress σ_{ef} shall be deduced from the acting local stress (Fig. 1.10). This effective stress is defined so the be equivalent, with reference to the material strength, to the acting stress state following the proper failure criterion. The various failure criteria have been elaborated for the different materials through experimental investigations and related theoretical interpretations.

In general, the overtaking of the maximum stress capacity in a point of the section does not take it to the ultimate limit of its resistance. The internal force can increase further. Therefore, the verification of the local stress is in a certain measure conservative that is at the safe side.

Following the second verification mode, at one side from the acting loads the distribution of internal forces is defined through the analysis of the structure; at the other side from the strength of materials, the resistance of the sections is deduced. For any section, the comparison between the acting and the resistant force is then made. To be noted that, for this *verification of the internal force of the section*, its ultimate limit state shall be defined on the basis of the proper failure criteria. These criteria are based mainly on the material deformability limits.

Also, the internal force is a quantity expressed by different scalar components (e.g. axial force and bending moment). The comparison between resistance and action shall be properly set up, for example defining, in the co-ordinated field of the different components, the failure borders that separate the resistance domain from the failure one (Fig. 1.11). For the verification, the point corresponding to the co-ordinates of the acting force shall be included in the internal resistance domain.

The failure border is constituted by the points corresponding to the ultimate resistance forces of the section. If one only component is dealt with (e.g. uniaxial bending), the border becomes a point and the comparison is made between scalar quantities. If two components are deal with (e.g. axial force and uniaxial bending), the border is a curved line and the comparison can be represented graphically in the plane as

Fig. 1.11 Example of failure border for combined axial force and bending moment

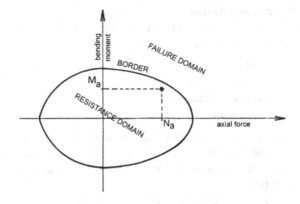

indicated in Fig. 1.11. If three components are dealt with (e.g. axial force and biaxial bending), the border is a curved surface and the comparison is to be performed in a three-dimensional space, possibly using proper graphic plane representations.

In general, the overtaking of the maximum force capacity in a section does not take the structure to the ultimate limit of its resistance. The applied loads can increase further. Also, the verification of the internal force of a section is in a certain measure conservative that is at the safe side.

The more exhaustive verification is given by the third mode that however for the moment presents relevant calculation difficulties and heavy operational disadvantages. It consists of the deduction of the force capacity of the sections from the material strengths and from these capacities of the bearing capacity of the structure. This bearing capacity is expressed in terms of a resistant load to which the acting load shall be compared.

Since in general different contemporary load units are acting on a structure, such calculation shall be repeated "for all the possible load combinations", unless conventional procedures are applied that fix the ratios among the intensities of the loads. So for instance, with reference to the nominal value F_i of the loads taken from the competent regulation, if one assumes for all the contemporary loads an increase proportional to one only factor λ, the ultimate limit collapse situation of the structure is defined by one scalar collapse factor λ_r that can be directly compared to the admissible value with $\lambda_r < \overline{\lambda}$. But not always, within the respective variation fields, the contemporary proportional increase $\lambda_r F_i$ of any single load leads to the "nearest" collapse threshold.

The *verification of the bearing capacity of the structure* can be performed following some calculation methods, like the "Plastic Hinges Method", set up with specific reference to the material type and the particular behaviour of sections and members. The applicability limits of these simplified methods are narrow, while in the general case a complex analysis of the structure is required to follow its behaviour beyond the field of linear elasticity. This does not consider the very simple cases, like for instance the isostatic arrangements that can be analysed by the basic equations of the statics.

Calculation Model

The paths indicated in Fig. 1.9 are run analytically on a proper model that interprets the real structure in an idealised way. The *calculation model* consists of:

- constitutive laws;
- geometric assumptions;
- definition of limit states;
- static scheme.

The *constitutive laws* refer to the material behaviour and are deduced from the measurement of forces and displacements with tension or compression tests. The results are translated into σ-ε diagrams of which in Fig. 1.12 some typical examples are shown.

The diagram of Fig. 1.12a refers to an elastic material with brittle failure (e.g. the glass). The diagram of Fig. 1.12b refers to an elastic material with ductile failure (e.g. with good approximation the steel). The diagram of Fig. 1.12c refers to an elastic–plastic material with brittle failure (e.g. with good approximation the concrete). The diagram of Fig. 1.12d refers to an elastic–plastic material with ductile failure. Deeper details are given in the specific texts of the single construction materials. This text recalls only some parameters that affect the structural behaviour, such as elasticity, plasticity, brittleness and ductility. The integration of the constitutive laws over the volume of the structure gives its behaviour that is its response, in terms of stresses and deformations, to an applied load distribution. With proper *geometric assumptions,* the problem can be much simplified with respect to the general three-dimensional one, adapting the algorithms to the specific categories of elements. So, one can refer to the *plates,* with one dimension small with respect to the other two, the behaviour of which can be related to the elastic surface representing the deformation of their mid-plane. One can refer to the *beams,* with two dimensions small with respect to the third one, the behaviour of which is related to the elastic line representing the deformation of their central axis. In the model, these categories are assumed

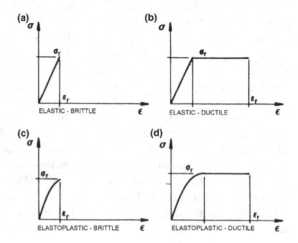

Fig. 1.12 Typical σ − ε diagrams

as elements with null transverse dimensions and with mutual joints consequently reduced. The competent integrations are extended to a surface or a line, being the transverse behaviour implicitly defined by the basic *Bernoulli's assumption* of the section planarity.

Again about the geometric assumptions, the one referring to the deformation magnitude is to be mentioned, following which a *first-order theory* that neglects the flexural displacements due to loads is to be distinguished from a *second-order theory* that does not neglect them.

After the analysis of the structure, the evaluation of resistance depends on the definition of the *ultimate limit state*. This definition is established with reference to the possible causes of structural collapse as listed here below.

1. *Loss of equilibrium as rigid body*: it is analysed by means of the basic equations of the statics in terms of resistant load (third mode of verification).
2. *Localised failure due to static actions*: it refers in general to single sections of the elements and is analysis in terms of resistant force (second mode of verification).
3. *Collapse due to transformation in mechanism*: it refers to the whole structure or a part of it of which the plasticization of some sections leads to a hypostatic situation with the attainment of the collapse load (third mode of verification).
4. *Collapse due to instability*: it refers to the possibility of instable deflection of the structure or a part of it due to the attainment of the "critical load" (third mode of verification).
5. *Localised failure due to fatigue*: it is produced by the repetition of alternate load cycles that leads to local brittle failure of the material due to peak stresses (first mode of verification).
6. *Other geometric disarrangements*: possibly deriving from excessive static or dynamic deformations incompatible with the service of the construction.
7. *Physic or chemical degradation*: it refers to the structural integrity jeopardised by the possible effects of the ambient actions, not always linked to the loads and sometimes linked to the duration properties of the materials.

The first four types of collapse are shown in Fig. 1.13. The fifth mode of the fatigue refers mainly to the moving parts of the machines; in the civil constructions, such as industrial buildings and road infrastructures, it refers to the lifting means (rails for bridge cranes), the transit of vehicles (road and railway bridges) and the other cases of structures subjected to a relevant number of cyclic load repetitions. The geometric disarrangements require the calculation of the deformations (e.g. the deflections of beams and floors) or the vibrations (e.g. versus the danger of resonance with the supported machines). For the physic or chemical degradation, proper technological prescriptions are needed related mainly to the material properties and to the adequate protections, but also real structural calculations are applied such as those that limit the stress level in-service conditions.

The *static scheme* of the structure is constituted by a graphic representation of its arrangement with elements, joints, supports, loads, dimensions, etc. (see for example Fig. 1.14). The static scheme can also include, when relevant for safety, the construction tolerances. These represent the variation field of the measures around

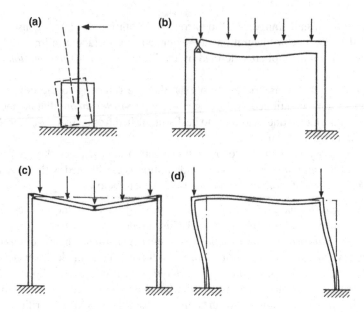

Fig. 1.13 Representation of failure modes

the nominal values indicated in the project. Regarding the structures of the concerned constructions, the tolerances can refer to the setting out, the positioning, the verticality, the horizontality, the length, the thickness, the straightness, the planarity, etc. In order to take into account them in the structural calculations, the nominal measures have to be modified in the most unfavourable way in the limits of the prescribed tolerances.

The reliability of the calculation model depends on the degree of correspondence between its behaviour and the real behaviour of the structure as "measurable" in situ. In general, the reliability increases with the greater complexity of the model and by consequence with the greater calculation burden. It depends also on the specific structural situations, for example isostatic or hyperstatic arrangements, that account for certain simplifying assumptions.

By means of the calculation model, the paths of Fig. 1.9 are performed: in the left direction that is from the applied loads to their effects in terms of internal forces or stresses, in the right direction that is from the material strength to the sectional resistances or the bearing capacity of the structure. The former paths lead to linear relations if, with negligible or not affecting displacements:

– the situation is isostatic so that the only basic equations of the statics are sufficient;
– the situation is hyperstatic, but the material is perfectly elastic.

The latter lead to linear relations under the same eventualities and if the material has a perfectly brittle failure.

Fig. 1.14 Example of a
static scheme of a structure

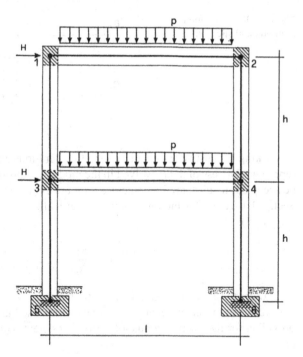

Only in the cases cited above, the ultimate resistance limit evaluated on the basis of the local stress coincides with those of the sectional force and the load on the structure and the three verification modes remain equivalent. In all the other cases, that include the second-order behaviour with large displacements and, for the hyperstatic structures, the non-elastic materials and the ductile failures, the safety evaluated on the structure does not coincide with the safety evaluated on the section or on the local point.

1.1.3 Semi-probabilistic Method

The probabilistic characteristics of the involved quantities, as already said, are not deduced directly from the structural assembly. They are tested at one side on some samples of the elementary materials; at the other side, the loads are assumed on the basis of investigations on similar situations. These data are introduced in the calculation model on which finally the necessary analyses are elaborated for the safety verifications. The probabilistic characteristics of data alter sensibly along the path, while new sources of errors come from the not perfect reliability of the model. For these reasons, the probabilistic criteria for safety verification shall be properly adapted.

Fig. 1.15 Representation of semi-probabilistic method

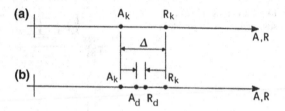

Lacking exhaustive information on these alterations and the precise probabilistic representativeness of data involved in the model, a semi-probabilistic method has to be accepted. It substantially coincides with the extreme values method presented in Sect. 1.1.1 except for the introduction of an additional safety margin. So, one sets (Fig. 1.15):

$$\gamma = \frac{R_k}{A_k} > \overline{\gamma}$$

In order to refine this equation, that still has partially the character of a deterministic verification, the coefficient γ can be decomposed in its three principal factors, providing for their separate quantification consistent with the uncertainties of each one:

$$\overline{\gamma} = \gamma_m \gamma_o \gamma_f$$

As indicatively illustrated in Fig. 1.16, the *Material Factor* γ_m takes into account the approximate correspondence between the materials of the tested samples and those actually realised in the construction; the *Action Factor* γ_f takes into account the approximate correspondence between the actions estimated with the initial investigation and those that actually can be applied on the construction; finally, the *Model Factor* γ_o takes into account the approximate reliability of the calculation model.

With such decomposition of the safety factor, the verification can be set as

$$\frac{R_d}{A_d} > \gamma_o$$

with

$$R_d = R_k / \gamma_m$$

Design Strength deduced from the characteristic strength with a reduction by γ_m and with

$$A_d = A_k \gamma_f$$

Fig. 1.16 Pertinent
application of safety factors

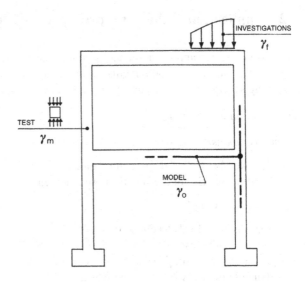

Design Action deduced from the characteristic action with an increase by γ_f.

The practical application of this verification method is better specified within the specific treatises of the different construction materials (concrete, steel, masonry, timber, etc.). For example, it will be shown how the material factors γ_{mi} can be properly differentiated for the different resistance components. In any case, presently the codes do not provide the quantification of the model reliability and, setting $\gamma_o = 1$, they lay the necessary safety margins on the other factors (γ_f. in particular). This still leaves sensible deficiencies since, moving from the simplest structures to the most complex, from the most drastic simplifications to the most refined representations, the calculation models can have much different levels of reliability.

Finally, it is reminded that the discourse presented up to here is principally referred to the safety verification of constructions. The design quantities A_d and R_d defined above represent, unless the not strictly probabilistic approximations of the partial factors γ_m and γ_f, the "limit state" characterised by a sufficiently small theoretical probability of being overtaken during the service life of the structure. More specifically, depending on the different purposes of the calculation, the structural analysis may refer to the evaluation of the following states:

- *ultimate limit state,* that starts from the most penalised *design values* of the involved quantities for the safety verification against collapse;
- *serviceability limit state,* that starts from the precautionary *characteristic values* of the involved quantities for the functionality verifications (deformations, vibrations, durability, cracking, etc.);
- *checking state,* that starts from the *most probable values* of the quantities or from the medium ones as measured for the prevision of the structural behaviour in the load tests or in general for the experimental in situ checks.

1.2 Application of the Semi-probabilistic Method

The actions that are applied to the constructions can be classified following different criteria. The first criterion distinguishes the nature of the actions as shown by the following scheme where the current-related symbols are also given.

Forces (or direct actions)

Permanent loads "G"

- Self-weight "G_1" of the structural elements;
- Dead loads "G_2" of the superimposed elements.

Variable actions "Q"

- Service (live) loads, snow load;
- kinetic wind pressure;
- thrust of soil (earth pressure) and loose materials;
- dynamic forces (vibratory, impulsive, etc.)

Applied deformations (or indirect actions)

- internal mutual action (prestressing "P", etc.);
- permanent distortions "G_ε" (subsidings, shrinkage, etc.);
- variable distortions "Q_ε" (thermic variations, etc.);
- earthquakes (ondulatory and auscultatory accelerations)

Aggressive conditions (physical or chemical)

- freeze/thraw, abrasion, etc.;
- humidity, chloride, saltiness, etc.

Finally, the *exceptional actions* "Q_{ex}" are defined, such as collisions, explosions, typhoons, etc. For these actions, usually an explicit resistance verification is not required, but only proper measures are prescribed able to limit the effects of possible failures.

Other classification criteria distinguish the actions on the basis of the different characteristics that can affect the structural analysis, such as those listed below.

MECHANICAL EFFECTS ON THE STRUCTURES

- static actions;
- dynamic increments;
- dynamic actions.

The first ones do not cause sensible accelerations and can be treated with a static analysis; the dynamic effects of many types of actions can be treated again with a static analysis having properly increased their values (see for example the "dynamic factors" related to the running loads on bridges); other actions require a dynamic analysis of the structure (e.g. structures supporting engines with rotating masses).

Possible range of the load intensity

- determinate action;
- limited action;
- alternate action.

The first one is related to one only value of its intensity (such as the self-weight of the construction elements); the second one can be present with any value of its intensity within zero and its possible maximum (such as the service loads); for the alternate actions finally the direction reversal is possible (e.g. the wind pressure).

SPACE VARIATION OF THE POSITION

- fixed loads;
- free loads;
- moving loads.

The first ones are related to only one possible arrangement on the structure; the second ones are related to different possible arrangements and to different possible combinations of the structural effects; the mobile ones refer finally to the running loads such as the vehicles that transit on the bridge decks or the bridge cranes that move on the supporting beams in the industrial buildings.

TIME VARIABILITY OF THE INTENSITY

- constant action;
- continuous action;
- incidental action.

The first one is permanently present without sensible variations (such as the weight of the structural elements); the second one has again a continuity feature, but with possible long-term variations (such as the weight of some completion works); the incidental actions represent a discrete phenomenon with possible repeated load events of short duration with respect to the service life of the construction (such as the crowding of the rooms or the wind gusts).

PROBABILISTIC TIME FEATURES

- invariable phenomenon;
- stationary phenomenon;
- transitory phenomenon.

Where in the first case, the probabilistic features of the intensity does not depend on the time; in the second case, they vary along the time with short recurrences with respect to the service life of the construction so that final representative values can be defined; if on the contrary the recurrences of the random variations have a return period longer than the service life of the construction, then the probabilistic values become dependent from the duration itself (such as for some natural phenomena of snow, floods, earthquakes, etc.).

MEAN PER CENT DURATION

- permanent loads;
- frequent load;
- rare loads.

Where the related values come into play for example for some calculation that depend on the load duration, such as those related to creep and cracking effects on concrete structures, and are represented by proper fractiles of the time distribution of the intensity.

1.2.1 Representative Probabilistic Parameters

Some short notes on probability calculation are here reported for what concerns the representative values of the different types of action, repeating that at present the possibility of a complete and rigorous application of the probabilistic criteria is limited also if referred only to the loads.

So, interpreting the intensity x of the generic action as a random quantity, as already presented in Sect. 1.1.1, its complete description can be set in terms of *probability density*

$$f(x)$$

that gives the elementary probability with

$$f(x)dx = P\{x < \xi < x + dx\}$$

or in terms of its *distribution function*

$$F(x) = \int_{-\infty}^{x} f(\xi)d\xi$$

that gives directly the progressive (or cumulative) probability

$$F(x) = P\{\xi < x\}$$

where the first function can be deduced from the second one with

$$f(x) = \frac{dF(x)}{dx}$$

With reference to the given distribution, some "central values" of the random quantity are defined. The *mean value* is

$$\overline{x} = \int\limits_{-\infty}^{+\infty} x f(x) dx$$

The *most probable value* x_o corresponds to

$$f(x_o) = \max$$

The *median value* x_m that separates the two parts of equal probability corresponds to

$$F(x_m) = 0.5$$

The three central values, that in general are different, can coincide for some types of symmetric distributions. If the random quantity x is discrete and can have a number N of values, then the mean value of the whole population $x_1, x_2, \dots x_N$ is given by

$$\overline{x} = \frac{1}{N} \sum_{i=1}^{N} x_i$$

The parameter linked to the dispersion of the values is the *variance* defined as

$$v = \int\limits_{-\infty}^{+\infty} (x - \overline{x})^2 f(x) dx$$

for a continuous random quantity x, defined as

$$v = \frac{1}{N} \sum_{i=1}^{N} (x_i - \overline{x})^2$$

for a discrete random quantity x_i. Usually, the square root of the variance is used. It has the same dimension of x (e.g. of the load intensity of the concerned problem). This latter parameter

$$s = \sqrt{v}$$

is called *standard deviation* and represents, with a term of geometry, the radius of gyration of a mass distributed like f(x).

If calculated on the basis of a sample of n values small with respect to the whole population of the possible values of x, one obtains an approximate mean value

$$\overline{x} = \frac{1}{n}\sum_{i=1}^{n} x_i$$

that does not coincide with the real one. For this approximate "partial" representation, an overestimation of the deviation is usually given with

$$s = \sqrt{\frac{\sum_{i=1}^{n}(x_i - \overline{x})^2}{n-1}}$$

This overestimation trends to fade out as the number n of values increases.

Assuming that the mean value gives the magnitude order of the expected values of the quantity x, the *variation factor*

$$\delta = \frac{s}{\overline{x}}$$

is meaningful, representing the ratio between the standard deviation and the mean value itself. One can say for example that a certain quantity can be determined with a good precision if a small variation factor shows a limited relative random dispersion.

In many cases, the structural analysis refers to the *characteristic values* of the loads, taken as limits with sufficiently small probability to be overtaken. For a random quantity x, the characteristic value x_k with probability $\overline{p}$ is the one that corresponds to the fractile

$$F(x_k) = \overline{p}$$

Assuming for the probability "no" a small limit (e.g. $\overline{p} = 0.05$ as set by many codes), by consequence one can distinguish two characteristic values related to the ends of the interval "yes": the lower x_k' that, as said above, corresponds to a probability $\overline{p}$ to be overtaken downwards, the upper one x_k'' that at the other side corresponds to the same probability to be overtaken upwards. For this latter, one has

$$F(x_k'') = \overline{q}$$

with

$$\overline{q} = 1 - \overline{p}\,(= 0.95)$$

The characteristic values can be evaluated with

$$x'_k = \bar{x} - k's$$
$$x''_k = \bar{x} + k''s$$

where $\bar{x}$ and s are the mean value and the standard deviation already defined and the factors

$$k' = k'(\bar{p})$$
$$k'' = k''(\bar{q})$$

depend from the shape of the density function $f(x)$ and have the same value for symmetric distributions.

Load Duration

When, with reference to a variable load, the duration of the different intensity levels is concerned, a special statistics is to be elaborated measuring, for any "unit duration" Δt the correspondent intensity level. The measurements have to be extended for a sufficiently long-time period usually coinciding with a functional cycle. For example, one can measure every hour the load intensity for a whole year.

In this way, it is possible to draw a histogram of the frequency density, laying out steps of height proportional to the number of hours/year in which the load has acting with a level x of intensity between x_i and $x_i + \Delta x$. The measurements can be limited to a meaningful reduced sample of hours distributed along the year. Also for this histogram, a continuous analytical model $f(x)$ can be formulated.

The mean $\bar{x}$ of such time distribution of the intensity defines the *quasi-permanent value*

$$x_q = \bar{x}$$

while the *frequent value* x_f is the characteristic one corresponding to the fractile $\bar{q}$ (=0.95)

$$F(x_f) = \bar{q}$$

Higher *rare values* can occur with very short duration for the incidental actions.

Load Peaks

In order to foretell the maximum intensity of the load that will be applied to the structure, one has to measure the peak values of the subsequent load events, without regarding their duration. Usually, incidental actions are concerned, such as the wind gusts, but also the continuous action can be involved, such the ambient thermal variations or the snow load at the different altitudes and latitudes. Of the intensity, only the maximum values can be measured, as for the snow load, or also the minimum values, as for the temperature.

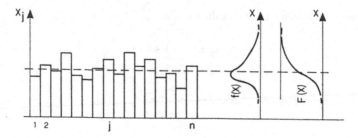

Fig. 1.17 Distribution of maximum values

So, let's consider the load peaks measured along the year j:

$$x_{1j}, \; x_{2j}, \; \ldots \; , \; x_{ij}, \ldots$$

The time distribution of the peaks has, for the climatic–meteorological actions, a typical shape; but this seasonal recurrence does not come into play, because only its maximum value is of interest. Nor one can deduce, from this seasonal distribution, information about the maximum values that may occur in the following years.

Prolonging the measurements for a number n of years, one can take for every year its maximum value X_j and elaborate the available set

$$X_1, \; X_2, \; \ldots \; , \; X_n$$

with the criteria of the *Theory of the extremes*. On the basis of this theory, from the set of the annual maximum values (see Fig. 1.17) one obtains the frequency density $f(X)$ that describes their statistical distribution. The correspondent distribution function provides the probability that the maximum X_j of a single year does not overcome a given level X. That is one has

$$P\{X_j < X\} = F(X)$$

and in a complementary way

$$P\{X_j > X\} = 1 - F(X)$$

With "repeated trials" for m subsequent years, from the product theorem one obtains the probability

$$P_m\{X_j < X\} = [F(X)]^m$$

that all the m annual maximum values remain lower than the given level X or, in a complementary way, the probability

$$P_m\{X_j > X\} = 1 - [F(X)]^m$$

that at least one of the cited m annual maximum values overtakes the same level X.

Following an alternative representation mode, any level X of the intensity can be related to an *average return period* $\overline{T}(X)$ that, omitting here the relative demonstration, is expressed in terms of progressive probability by the (approximate) relation

$$\overline{T}(X) = \frac{1}{1 - F(X)}$$

In the opposite direction, the probability is set as a function of the average return period with

$$P\{X_j < X\} = 1 - \frac{1}{\overline{T}(X)}$$

$$P\{X_j > X\} = \frac{1}{\overline{T}(X)}$$

Finally, for a number m of subsequent years the expressions of the probability can be approximated to an exponential form

$$P_m\{X_j < X\} = \left[1 - \frac{1}{\overline{T}(X)}\right]^m = e^{-m/\overline{T}(X)}$$

$$P_m\{X_j > X\} = 1 - \left[1 - \frac{1}{\overline{T}(X)}\right]^m = 1 - e^{-m/\overline{T}(X)}$$

valid for average return periods sufficiently long ($\overline{T} > 10$ years).

In a practical problem of structural analysis, m is the expected (or design) life of the construction. Having assumed a probability $\overline{p}$ sufficiently small that, during this life, the maximum annual load X_j overcomes the level X to be used in the analysis, by consequence one obtains, from the last relation written above, the correspondent average return period

$$\overline{T} = \frac{m}{\ln[1/(1 - \overline{p})]} \simeq \frac{m}{\overline{p}}$$

again expressed in years like the life m.

Finally, knowing the distribution function $F(X)$ of the analysed phenomenon, the following equation is set

$$\frac{1}{1 - F(X)} = \overline{T}$$

from which one obtains the *reference value* X_T. For the solution of this equation, one can use specific probability charts with tables and graphics.

Fig. 1.18 Normal Gauss
form

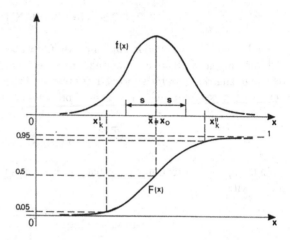

The algorithms of the theory of extremes, as presented here above, are well applied to phenomena of which the biggest events have long recurrences, measurable in tenths of years. That is for average return periods of the same order or also longer than the expected life of the construction. For short return periods, for example day or week recurrences, as those off many service loads, these algorithms have to be improved.

For these latter types of loads, one can also follow an alternative simplified way. If the peak frequency is high and the characteristics of the statistical distribution computed on samples more and more extended (one week, one month, four months, one year, five years, etc.) shows a stable stationarity of the phenomenon, with mean value, standard deviation and extreme value practically unchanged, then of this distribution one can assume directly the characteristic (upper) value x_k, the one for which it is

$$F(x_k) = \bar{q}$$

as for the invariable phenomena with "topological" and not time randomness. This means that along the life of the construction, bigger events are not expected, since they would be out of the interval of possible random variation as finally delimited by the preceding experimentation.

Models of Statistic Distribution
To conclude, here below the models of statistical distribution mostly employed are shown. The first one assumes a symmetric function f(x) with the *normal (or Gauss) form* (Fig. 1.18):

$$f(x) = \frac{1}{\sqrt{2\pi}s}e^{-\frac{(x-\bar{x})^2}{2s^2}}$$

Fig. 1.19 Definition of the reliability index β

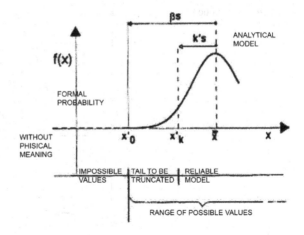

that is defined on the basis of the mean value $\bar{x}$ and the standard deviation s. For $\bar{p} = 0.05$, one has $k = k' = k'' = 1.645$ and the two characteristic values are given by

$$x'_k = \bar{x} - 1.645s$$
$$x''_k = \bar{x} + 1.645s$$

The normal form described above is mainly referred to the topological randomness of some mechanical properties of materials such as the weight. It gives small but not null probabilities also for values of the random quantity very far from the mean one. Actually out of a certain range, there cannot be experimental values unless for systematic events that do not refer to the random variability of the quantity of concern. The not null probability given by the analytical model out that range are impossible or even without physical meaning as for the negative values of the quantity. So, referring for example to the material strength, the tail under the lower characteristic value should be truncated at the limit value

$$x_d = \bar{x} - \beta\,s$$

where β is the *reliability index* quantified on the experience basis (see Fig. 1.19).

Instead of the normal Gauss form, which is defined for $-\infty < x < +\infty$, one may adopt the *log-normal form* which has the same analytical model of the normal form where it is set $\xi = \ln(x - x_d)$:

$$f(x) = \frac{1}{(x - x_d)\sqrt{2\pi}\sigma}e^{-\frac{(\xi - \bar{\xi})^2}{2\sigma^2}}$$

where $\bar{\xi}$ is the mean value of ξ and σ is its standard deviation. This latter asymmetric form, defined for $x_d < x < +\infty$, can be used for the probabilistic definition of the lower values of the material strength.

Fig. 1.20 Asymptotic first
type form

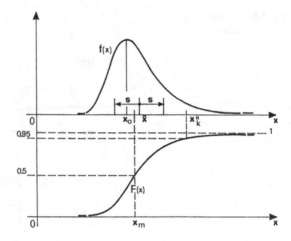

Another model assumes a function f(x) with the fixed dissymmetry of the *asymptotic* first *typed form* (Fig. 1.20) derived from

$$F(x) = e^{-e^{-\alpha(x-u)}}$$

The parameters of this expression can be evaluated, with reference to a sufficiently numerous sample, as a function of the mean value $\bar{x}$ and the standard deviation s with

$$\alpha = 1.28/s$$
$$u = \bar{x} - 0.45\,s$$

The first parameter is inversely linked to the dispersion of the statistical distribution; the second one represents the most probable value ($u = x_0$). To obtain the upper characteristic value (on the fractile $\bar{q} = 0.95$), one sets

$$x_k = \bar{x} + 1866\,s$$

while for the reference value (with return period $\bar{T} = 50$ years), one sets

$$X_T = \bar{x} + 2.608\,s \quad (\bar{q} = 0.98)$$

The asymptotic form described above is employed, mainly for time variabilities, on three levels: the *integral continuous* one that measures for any load level the correspondent duration (for the definition of the frequent and quasi-permanent values); the *discrete peak* one that measures all load peaks without reference to their duration (for the definition of the characteristic stationary value); the one of the *extreme values* only, that for example measures the maximum annual loads (for the definition of the reference value with a given return period).

To conclude, it is to be noted that, when the correct measurements of the probabilistic parameters are not available, the load is necessarily represented by *indicative values* deduced from the current design praxis on the basis of a long applicative experience. In particular, the *nominal values* are cited that, among the indicative ones, have the "authority" given them by the specific technical regulations.

1.2.2 Action Analysis

For the *weights of materials* that can be involved in the calculation of the actions on the constructions, there are exhaustive statistical investigations. These show random distributions of the values well represented by the normal form presented above. Reference is made to the topological randomness of quantities that are constant in the time.

In the specific regulations, one can find wide tables that give the mean values (or the upper characteristic ones) with the related standard deviations (or the variation factors). So with $\bar{p} = 0.05$, for the structural analysis two possible values should be assumed: the upper characteristic one

$$g_k'' = \bar{g} + 1.645s$$

if the weight g is unfavourable for safety, the lower characteristic one

$$g_k'' = \bar{g} - 1.645\,s$$

if the weight is favourable for safety.

About these definitions, that appear conceptually correct, the following comments can be made.

- The statistics on the specific weight of the materials show small dispersions of the values ($\delta = 0.5$–1.0% for metallic materials, $\delta = 5$–10% for stone materials, $\delta \approx 10\%$ for timber).
- The values of actions are also influenced by the execution tolerances of the works, tolerances that vary with the type and the dimension of the elements (going from negligible fractions for metallic or massive elements to levels also greater than 20% for slender and thin elements of stone materials).
- For some composite construction materials, the volumetric weight depends also from the stress level (as for the reinforced concrete that can have incidences of steel reinforcement very different).
- While for the resisting structures it is possible to define with precision since the design stage the geometric characteristics related to the computation of the self-weight, this is not possible for the dead loads of the finishing works that are often foretold in a general way and can be modified along the life of the construction.

For these considerations, it is reasonable to adopt the approximation of the *technical criterion* that interprets as determinate the quantity G related to all permanent loads, taking it from one only value (e.g. from the mean one) of the volumetric weights and from the "nominal" dimensions of the works, and moving into the pertinent partial safety factor γ_g the randomness related to the intrinsic material non-homogeneity, the execution tolerances and the possible design differences.

Service Load

For the service loads of the buildings, specific tables are available that give, for the principal types of civil destinations, the representative values of the statistic distributions. About this type of variable actions, one has to consider the composite nature of the time variability (Fig. 1.21): the part related to the furniture (and basic equipment) is of continuous type with possible long-term recurrent variations; the weights of wares can have similar features; incidental actions are added for crowding that has in general much more frequent recurrences.

For this type of time variability with short recurrences, with a stationary interpretation it is possible to elaborate the whole set of peaks taken as completed within the construction life. One can set for example 50 years for this service life, a proper representative reduced sample is taken, and on this, the parameters necessary for the definition of the probability function are evaluated. This function, for the service loads on civil buildings, seems to be well represented by the asymptotic first type form described in a previous section. The designer, for the resistance calculations, assumes of it the upper characteristic value q_k ($=q_k^{''}$).

A relevant problem arises in foretelling with certainty the type of functional use of the rooms. The more this use is delimited, the more reliable is the definition of the related actions. On the contrary, the designer usually has to refer to wider types of destinations, correspondent to the urbanistic class of the building permit.

To clarify the question, let's consider the example of classification of the following list. For any class of destination, an indicative value of the service load is given (round figure) and set as the bigger characteristic value of the different categories of activities specified in the last column.

Fig. 1.21 Time variability of service loads

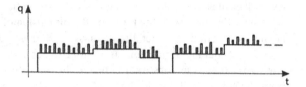

DESTINATION CLASSES OF CIVIL BUILDINGS	Service loads q_k (kN/m^2)	
Dwelling and similar	2.0	Dwellings
		Hotels
		Offices
		Services
Open to public	3.5	Shops
		Offices
		Restaurants
		Schoolrooms
With crowding	5.0	Halls
		Theatres
		Churches
Special	≥ 6.0	Dancing
		Stadiums
		Archives
		Libraries

To give an idea of the possible differences of the service loads within a class, the reduction factors of the indicative value of the class itself for some types of the category "shops" are reported here under.

Libraries	0.85
Ironmongeries	0.85
Stores	0.71
Chemists	0.50
Wear	0.50
Furniture	0.36

The statistical investigations on the service loads on buildings are referred to a *minimum representative surface* of the floor and give the intensity values (kN/m^2) to be taken as constant over this area for the purpose of the design of the deck elements (floors and beams). Reference can be made to the mean value over 10 m^2 of floor. If referred to larger surfaces, the mean loads show values that decrease with their area. In some specific standards, the load reduction formulae can be found with reference to the different categories of destination. Starting from the maximum indicative one over 10 m^2, a value decreasing linearly dawn to the minimum for surfaces equal or greater than 100 m^2 can be assumed.

In a simpler way, the design praxis follows often the criterion of assuming two values for the service load: a bigger one for the local verifications of the floor elements; a reduced one for the global verifications of the whole structural system. In addition concentrated actions may be assumed for possible particular verifications under the load prints.

What presented above shall be integrated into the combination rules of the different contemporary load units as further on specified.

Snow Load

The snow load represents the classic action of climatic–meteoric type for the evaluation of which the criteria of the theory of extremes can be well applied. Statistic data are available with sufficient numerousness, related to the different snow-metric stations spread over the territory. So, elaborating the annual maxima measured for a sufficiently long period, the mean values have been deduced, systematically differentiated on the basic of the altitude and the geographic position. For all the sites, the same value $\delta = 0.5$ of the variation factor seems to be valid.

Assuming the asymptotic first type model, for the different return periods one obtains the corresponding values q_s of the maximum annual level, as shown in the following table (with $\delta = 0.5$).

$\overline{T}$ (years)	$F \left(= 1 - 1/\overline{T}\right)$	$1 + k\delta (= q_s/\overline{q}_s)$	$\alpha_r (= q_s/q_{os})$
20	=0.950	=1.93	=0.84
50	=0.980	=2.30	=1
200	=0.995	=2.87	=1.25
1000	=0.999	=3.45	=1.50

Instead of the mean one $\overline{q}_s$, the standards usually give the reference value q_{os} of the snow load that corresponds to mean return period of 50 years. For different periods, the values have to be modified with the *return factor* α_r.

As already said, the mean return period is computed on the basis of the years m of the expected service life of the construction and of the risk level expressed by the probability $\overline{p}$. For $\overline{p} = 0.05$, assuming the 50 conventional years of life of the "permanent" constructions, one has to refer to $\overline{T} = m/\overline{p} = 1000$; for temporary constructions with an expected service life not longer than 10 years, one can refer to $\overline{T} = 200$; the period $\overline{T} = 50$ of the reference load instead corresponds to the provisional works or to the execution stages with duration m not greater than 2.5 years.

The reference load is conventionally defined on the basis of a subdivision of the territory in climatically homogeneous zones. This discretised classification of the snowfall conditions of the different regions has sensible approximations and leads to indicative values of the relative load. So, indicating with q_o the basic value related to the zone with the greater snowfalls (e.g.: $q_o = 0.9$ kN/m^2), for the other zones the snow load is computed with $q_{os} = \eta q_o$, where for example the reduction factor η can be set $= 2/3$ for the zones with medium snowfalls and $= 1/3$ for the zones with smaller snowfalls. The load q to be used in the structural analysis is evaluated with the addition of an altimetry function $\beta = \beta(h)$ that accounts for the site altitude h:

$$q = \beta \, \eta \, \alpha_r q_o$$

The designer has finally to evaluate the effects of the particular construction configuration, such as the reduction of the snow load on the roofs with high pitches and the snow accumulation in the compluvia or against possible restraints.

To this end, the pertinent standards give the values of special *exposition factor* μ with reference to some common types of roofs.

Wind Thrust

The *maximum wind speed* is sensibly independent from the local geographic conditions. At a height of 10 m from the ground, in a level zone without relevant obstacles, the maximum annual speed of the wind gust that has a minimum duration of 10 min keeps, all over the earth globe, around the mean value

$$\overline{V} = 22\,\text{m/s} \quad (=79\text{Km/h})$$

with a variation factor δ contained between 0.23 and 0.14. This does not refer to the catastrophic events such as typhoons that are interpreted as exceptional actions.

In order to deduce, from the one quoted above, the wind speed in the site, one has to take into account:

- the local topographic conditions of higher or lower exposition;
- the mean return period of the action in relation to the expected life of the construction and the assumed risk level;
- the rise position over the ground level and the surrounding obstacles.

So, the conventional procedure of evaluation assumes a *site speed* equal to

$$V_z = V_{ref}\, \alpha_l\, \alpha_r\, \alpha_z$$

and gives the criteria for the definition of the three factors.

The *topography factor* α_l refers to the local orographic conditions of higher or lower exposition with respect to the standard situation. The value $\alpha_l = 1.10$ can for example be assumed for sites on hills top or at gorges entrance; the value 0.90 can for example be assumed in sheltered dells.

The *return factor* α_r comes from a statistical evaluation of the anemometer measurements based on the asymptotic first type model as for the snow. For the speed of the wind gust, the dispersion of the maximum annual values is small. Actually, the phenomenon shows short recurrences and, with reference to the conventional life of constructions, appears close to the stationarity.

The following table gives the values similar to those presented for the snow:

$\overline{T}$(years)	$F(=1 - 1/\overline{T})$	$1 + k\delta(=q_s/\overline{q}_s)$	$\alpha_r(=q_s/q_{os})$
20	=0.950	=1.25	=0.93
50	=0.980	=1.35	=1
200	=0.995	=1.49	=1.11
1000	=0.999	=1.66	=1.23

where the longer return period refers to the permanent constructions, the preceding one refers to the temporary constructions, the reference one with $\alpha_r = 1$ refers to provisional construction situations and corresponds (with $\delta = 0.135$) to the wind speed

$$V_{ref} = 30\,m/s \quad (= 108\,Km/h)$$

The *profile factor* α_z accounts for the different slowdown degrees produced by the ground roughness to the wind motion. This effect decreases with the height over the ground level.

In order to evaluate this factor, different roughness categories can be conventionally defined, for example with:

1. open sea;
2. seaboard and lowlands (reference category);
3. suburban and woodlands;
4. urban and mountainous.

To these categories correspond as many vertical profiles of the wind speed as shown for example in Fig. 1.22.

In common situations, the effects of the wind can be computed with a static analysis of the structure subject to the action deriving from the *kinetic pressure*:

$$p = V_z^2/1.6 \quad (N/m^2)$$

Fig. 1.22 Vertical profiles of the wind speed

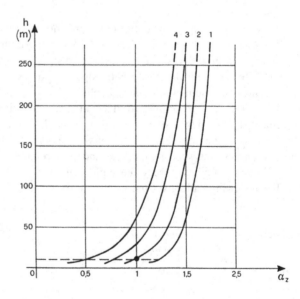

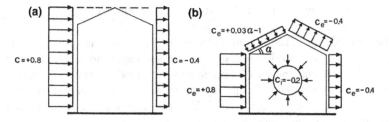

Fig. 1.23 Distribution of wind pressure for the calculation of: **a** overall stability, **b** local effects

As deduced on the basis of the criteria of the fluid mechanics, the kinetic pressure shall be corrected with

$$w = p\,G\,C$$

to obtain finally the *wind thrust* to be applied on the static scheme of the structure.

The *Gust factor G* refers to the variability of the kinetic pressure in relation to the features of the construction and can be numerically elaborated as a function of a *dynamic factor* α_d. This factor accounts for the reduction effects due to the non-contemporaneity of the maximum pressures on the surface extension struck by the wind and the amplification due to the vibration effects on the construction. In particular, these latter effects, little relevant in the rigid structures, are relevant in the flexible structures (e.g. high-rise buildings), for which a dynamic analysis can be necessary. The question, which is very wide, cannot be deepened here. Reference can be made to the specific documents. In the standards on actions, some approximate formulae are traceable together with some tables for the most common construction typologies.

Finally, the *exposure factor C* gives, for the different surfaces of the construction, the pertinent component of pressure or depression on which the wind thrust is decomposed. Two types of factors are distinguished: those related to the global behaviour of the construction, used in the stability calculations of the structural assembly with its bracing elements (Fig. 1.23a); those related to the local effects (external and internal), used for the resistance calculations of the single construction elements (Fig. 1.23b).

For example, for a building 20 m high, positioned in a zone of category 3 in an ordinary topographic area, one has:

$$V_z = 30 \times 1.0 \times 1.23 \times 0.92 = 33.95\,\text{m/s}$$
$$p = 33.95^2/1.6 = 720\,\text{N/m}^2$$

For the calculations of the overall stability of the building, with a Gust factor $G = 1.12$, one should assume a global wind thrust (upstream pressure plus downstream depression) equal to:

$$w = 720 \times 1.12 \times (0.8 + 0.4) = 968 \, \text{N/m}^2$$

More detailed information is available in the pertinent standards that give the exposure factors for many different situations. These factors are deduced from fluid dynamic analyses and special experimentations on models. The tests, performed in wind tunnel on reduced scale prototypes, are necessary for important buildings such as skyscrapers, suspension bridges, high-rise towers, etc., also with reference to possible resonance vibration phenomena.

1.2.3 Design Procedure

Summarising what before presented and using the symbols adopted in the Eurocodes, the "first-level" procedure for safety verification, corresponding to the extreme values method specified in Sect. 1.1.1, develops through the following three steps:

– The *actions* F_{kj}, expressed by their probabilistic representative values, are introduced in the static scheme of the structure, where the *materials properties* R_{ki} are also represented by their probabilistic representative values.
– On the calculation model, a deterministic structural analysis is then elaborated, with the nominal values of the concerned geometric dimensions, leading to the evaluation of the of the *effects* E in the structural members.
– The *safety verifications* are finally performed for all members, with the check of the safety fulfilment for the pertinent *limit states*.

This procedure, to which the name of *semi-probabilistic limit state method* is given, is the one presently proposed in the design codes.

The limit states considered for the verifications are:

  – *ultimate limit states* (ULS) corresponding to the structural failure;
  – *serviceability limit state* (SLS) for the construction functionality and durability.

For what concerns the former, the text will hereafter mainly refer to the resistance against the local failure of the structural members. For what concerns the latter, service limits will be considered for stresses in materials, cracking in concrete and deflection of floors and beams.

The verification with respect to the resistance ultimate limit state is obtained, applying the pertinent *partial factors*, with the comparison

$$R_d \geq E_d$$

where

– R_d is the design resistance calculated with the *design values* $R_{di} = R_{ki}/\gamma_{Mi}$ of the strength of materials.

– E_d is the design value of the effect of actions, calculated with the design values $F_{dj} = \gamma_{Fj} F_{kj}$ of actions.

The partial safety factors γ_{Mi} and γ_{Fj}, associated respectively with the *ith* material and *jth* action, cover the variability of the respective values together with to the incertitude relative to the geometric tolerances and the reliability of the design model.

The verifications with respect to the serviceability limit states are done at the level of characteristic values with

$$E_k \leq E_{lim}$$

where

– E_k is the value of the considered effect (stress in the material, crack opening or floor deflection) evaluated with the characteristic values of actions.
– E_{lim} is the corresponding limit value which guarantees the functionality of the building.

Combination of Actions

For *permanent loads G*, which have a small random variation, the mean value is assumed as representative. *The self-weight of the structure G_1*, that can be defined with higher precision at design stage, is distinguished from the *dead loads of non-structural elements G_2*, being these latter defined with lower precision.

Variable actions, such as imposed loads on floors, snow loads and wind thrust, are represented by their characteristic value Q_k. In order to account for the reduced probability that they act at the same time with their maximum values, the actions are scaled down in the combination formulae with the pertinent *combination factors*. These factors, with reference to the relative (per cent) duration of the different levels of intensity of the variable action, define the following combination values:

– *quasi-permanent $\psi_{2j}Q_{kj}$*: mean value of the time distribution of intensity;
– *frequent $\psi_{1j}Q_{kj}$*: 95% fractile value of the time distribution of intensity;
– *combination $\psi_{0j}Q_{kj}$*: value of small relative duration but still significant with respect to the possible concomitance with other variable actions.

For the different limit states' verifications, the following combinations of actions are defined.

– *Fundamental combination* used for ultimate limit states (ULS):

$$\gamma_{G1} G_1 + \gamma_{G2} G_2 + \gamma_{Q1} Q_{k1} + \gamma_{Q2} \psi_{02} Q_{k2} + \gamma_{Q3} \psi_{03} Q_{k3} + \ldots$$

– *Characteristic combination* used for irreversible serviceability limit states (SLS):

$$G_1 + G_2 + Q_{k1} + \psi_{02} Q_{k2} + \psi_{03} Q_{k3} + \ldots$$

– *Frequent combination* used for reversible serviceability limit state (SLS):

$$G_1 + G_2 + \psi_{11}Q_{k1} + \psi_{22}Q_{k2} + \psi_{23}Q_{k3} + \ldots$$

– *Quasi-permanent combination* used for the long-term effects (SLS):

$$G_1 + G_2 + \psi_{21}Q_{k1} + \psi_{22}Q_{k2} + \psi_{23}Q_{k3} + \ldots$$

In these formulae, "+" implies "to be combined with" and Q_{k1} represents the leading action for the concerned verification. Depending on the favourable or unfavourable effects for the verification, for the partial safety factors the following values can be taken, respectively:

Structural self-weight	$\gamma_{G1} = 1.0$ or 1.3
Superimposed dead loads	$\gamma_{G2} = 0.0$ or 1.5
Imposed loads	$\gamma_{Q} = 0.0$ or 1.5

What mentioned above refers to the verifications of the structure and foundation elements. For the verification of the foundation soil, one can refer to the specific chapter of the pertinent design codes.

The combination factors ψ_{1j} and ψ_{2j} refer to the per cent duration of the action and are given by the ratio, over the characteristic one Q_{jk}, of the correspondent values frequent x_f and quasi-permanent x_q defined in Sect. 1.2.1.

Also for the combination factors, proper values are proposed, as for example reported here under with reference to some categories of functional destination.

	ψ_0	ψ_1	ψ
Dwellings	0.60	0.35	0.20
Offices and shops	≥ 0.60	0.60	0.30
Garages	1.00	0.70	0.60
Snow and wind	0.75	0.20	0.00

For the multi-storey buildings with civil destinations (dwellings, offices, etc.), the calculation of the vertical structural elements can be performed reducing the service loads with the following combination factors ψ_o:

Roof	1.00
Last floor	1.00
Lower floors	0.70

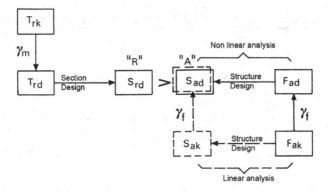

Fig. 1.24 Scheme of the calculation paths

Methods of Stress Analysis

The action amplification factors γ_{Fj} presented above should be applied to the loads that is at the origin of the paths that in Fig. 1.9 lead to the evaluation of their effects. With reference to the second verification mode, the one referred to the internal force of the sections, the right part of Fig. 1.24 indicates with the continuous line the rigorous path that first amplifies the action F and then performs the stress analysis evaluating the internal forces S in the sections (axial force, bending moments, shear forces, torsional moment). The left part of the same figure shows the opposite path that cuts down first, with their factors γ_{Mi}, the strength of materials and then evaluates the resistance capacities of the same sections. Between the acting internal forces and the correspondent resistance capacities, the safety verification comparison is finally set.

If the relations between the applied loads and the consequent internal forces are linear, as happens for example for the isostatic structures, the action amplification can be moved on, as indicated with the dashed path of Fig. 1.24. In this case, the result of the verification is the same.

On the contrary for the hyperstatic structures, the relation between the applied loads and the consequent internal forces depends on the constitutive law of materials. In general, this law maintains the structural behaviour in the elastic linear field up to the serviceability load levels and then deviates more and more from the linearity as the load approaches the ultimate failure limit. So, for these situations the lower path of analysis of Fig. 1.24, that leads from the characteristic value of action to its effect remaining within the elastic behaviour of the structure, can be performed with linear algorithms, the upper analysis that leads from the design value of the action to its effect needs nonlinear algorithms. In this case, the result of the verification at the end of the two quoted paths, the one that postpones the amplification and the correct one, is not the same.

Figure 1.25 shows examples of the two different situations discussed above. The case of the section A belongs to an isostatic system which is governed by the linear

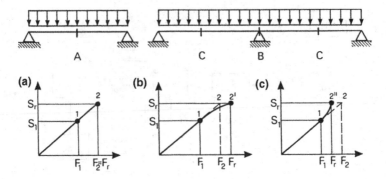

Fig. 1.25 Behaviour of sections **a** in an isostatic system, **b** and **c** in a hyperstatic system

relation represented in the diagram *(a)*: the safety factor evaluated on the internal forces S of the section is the same of the "true" one evaluated on the loads F:

$$\gamma = \frac{F_r}{F_1} = \frac{S_r}{S_1}$$

On the contrary, the cases of the sections B and C of Fig. 1.25 belong to a hyperstatic system for which diagrams of (b) and (c) type can turn out. In particular for the type (b), the real degree of structural safety, that is the one referred to the loads F, is greater than the one evaluated on the internal forces S:

$$\gamma = \frac{F_r}{F_1} > \frac{S_r}{S_1}$$

and the linear assumption would be wrong in the safe side, In the case (c), being

$$\gamma = \frac{F_r}{F_1} < \frac{S_r}{S_1},$$

it is the contrary and the linear assumption, with the amplification postponed on the internal forces, would be wrong in the unsafe side.

The discourse presented in this way is not complete. Type (b) behaviour belongs to the sections that, within a hyperstatic system, are the first to attain high levels of internal forces. Type (c) behaviour belongs to the sections that, within the same hypertatic system, are initially less stressed and then meet, with an increased engagement, the progressive yielding of the first ones. The ultimate load bearing capacity of the structure in general is limited by the first sections, and so the linear assumption applied for their verification is sufficient to guarantee the safety of the whole structural system. And this is true as long as in the force–deformation relation of the sections hardening behaviours, with stiffness increasing as approaching to failure, are excluded.

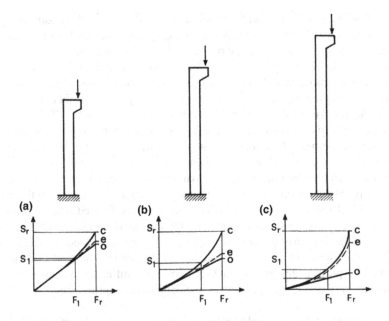

Fig. 1.26 Second-order effects in a column with **a** low, **b** medium and **c** high slenderness

For these reasons, within the applicative criteria of the limit states semi-probabilistic method, the design codes admit the simplification deriving from a *linear stress analysis*, which is much easier than the *nonlinear stress analysis* and saves the most important possibility of effects superposition.

Second-Order Effects

The linear simplification is not possible when, because of the presence in the structural system of slender compressed members, the second-order behaviour should be analysed with the algorithms presented the Chap. 4. In this case, to the mechanical nonlinearity of materials discussed here above, the geometric nonlinearity is added due to the sensible flexural displacements of the structural members.

While, for the beams of Fig. 1.25 lacking of second-order effect because of the absence of axial actions, the isostatic arrangement was associated with a perfectly linear behaviour, and in the hyperstatic arrangement, the nonlinearity was oriented so to help the most stressed sections, the (isostatic) column of Fig. 1.26 subjected to an eccentric axial compression has instead high nonlinear (second-order) effects that amplify the internal force in the critical section with swiftness increasing towards the collapse limit.

Figure 1.26 shows three situations of columns subjected to eccentric axial compression with different slenderness in order to underline the relevance of the effects. Reference is made to response curves expressed, for the simple columns of Fig. 1.26, in terms of the bending moment S_r in the most stressed section as a function of the applied load F. With "o", the target of a linear analysis performed under the critical collapse load F_r is indicated; with "e", the target of a second-order elastic analysis is indicated; with "c", the real intersection of the second-order analysis inclusive of the mechanical nonlinearity with the resistance limit S_r is indicated.

One can see in these diagrams that any linear approximation of the analysis leads to errors in the unsafe side. In the field (a) of columns with low slenderness, this error is small and some codes allow to neglect it as long as it does not overcome the 10%. In the field (b) of the medium slenderness, the second-order effect becomes important and is given mainly by the material nonlinear behaviour. In the field (c) of the high slenderness, the effect is even greater, with decreasing influence of the mechanical contribution and better precision of the nonlinear elastic analysis.

From what said above one can deduce that, except for the low slenderness, when instability problems are present a *complete nonlinear analysis,* inclusive of the material mechanical and the second-order effects, should be performed at the level of the design loads F_{ad} previously amplified. For the hyperstatic arrangements, as loads increase a redistribution of actions can rise helping the more stressed members when their flexibility increases due to second-order effects. But in general, it is not possible to know a priori if this hyperstatic help is or is not sufficient to compensate the opposite effect of accelerate stress increase. So, no linear approximation in any case is possible.

The instability problems could be solved by the definition of a critical load and the subsequent comparison of it with the load applied to the structure, following the third verification mode of Fig. 1.9. But this would be very difficult because of the multiplicity of the loads and their many possible different combinations. The procedure can be simplified if one moves the verification comparison to the load effects (internal forces) in the sections. But the problem remains rather complex. In fact if the properties of the materials cut down with the pertinent factor γ_M are used consistently with their strength design values, the internal forces analysis can be taken about the collapse limit where the solution is indeterminate.

Furthermore, the safety factors themselves have a well different meaning if referred only to the sections where the local resistance verification is performed or to the distributed properties of the current sections on which mainly depends the behaviour of the structural systems.

For these reasons, the following procedure can be proposed:

– At the side of the structural analysis, a factor γ_o connected to the model is resumed (e.g. $\gamma_o = 1.2$), taking in this way, with lower design loads, the algorithms off the indeterminateness field.
– For the material properties, a double definition is taken that differentiates the local resistance of the mostly stressed *critical sections* (with higher γ_M factors) from the distributed behaviour of the *current sections* (with lower γ_M factors).

Other approximate procedures are proposed to simplify the nonlinear analysis with specific reference to the different construction materials.

Structural Robustness

The discourse is to be completed with reference to some particular problems related to the safety of constructions. First, the eventuality of *chain collapse* is cited, following which the local failure of an element of a complex structure, caused by an exceptional event, has a chain propagation over a relevant part of the same structure. The importance of the problem has been pointed out by a catastrophic collapse that remained "famous" in the specialised literature: an explosion due to a gas leakage at a low floor produced the breaking down of a bearing wall panel and, with a chain progression, the collapse at full height of part of the residential tower of Ronan Point in London.

Against the chain collapse, some proper criteria shall be adopted. A correct conceptual choice of the overall construction is mainly needed, avoiding as much as possible the "serial" resistant arrangements (see for example Fig. 1.27a) and adopting independent "in parallel" arrangements (see for example Fig. 1.27b).

Where necessarily there are chains of elements working in series (as the superposed columns of Fig. 1.27c), the resistance of the base elements can be improved with a proper *behaviour factor* γ_o' (>1), the same with which one amplifies the horizontal actions on buildings when the bracing function, instead of being distributed on all members, are concentrated on single structural cores (Fig. 1.27d).

For some types of structures (e.g. wall panel systems), it is possible to perform a real *propagation design* by means of resistance verifications of the structure lacking of the collapsed element (Fig. 1.27e). For these situations, one can assume reduced

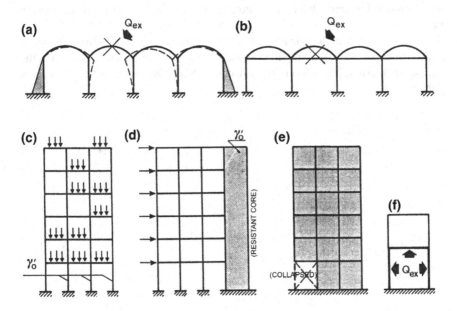

Fig. 1.27 Examples of chain collapses

safety margins (e.g. with $\gamma_F = 1$); their fulfilment requires in general to insert additional ties in the structural assembly.

In some cases of particular importance (e.g. bunkers for the storage of dangerous materials), a resistance verification can be performed against the exceptional action (Fig. 1.27f), using the *exceptional combination*

$$F_d = G + P + \psi_{31}Q_{1k} * \psi_{32}Q_{2k} + \ldots + \psi_{ex}Q_{ex}$$

where Q_{ex} is the representative value of the exceptional action, γ_{ex} ($= 1.0\text{--}1.5$) is partial factor, to be quantified in relation to the gravity of the damages consequent to the possible failure of the element of concern, ψ_{3j} are the combination factors referred to the load fractiles with duration equal at least to 30 days/year. This very heavy verification against the exceptional action concerns, not only the chain propagation of failure, but more in general the confinement of the damages of the exceptional event and by rules has to be accompanied by design provisions able to limit the intensity of the action (such as the vent openings of the burst overpressure).

Finally, it is to be reminded that, in addition to the resistance and serviceability verifications of the finished construction along its expected service life, all the preceding stages of its production process have to be checked. So, it is necessary to verify all the partial or transitory situations, including the possible provisional supporting works. The designer shall consider the different *transitory stages*, analysing the partial static schemes with related loads and materials of the moment. In particular for prefabricated elements, the verifications should follow the elements since from their factory production, through handling, storage, transportation and final installation in site.

The verifications in the transitory stages refer, further to the necessary integrity of the element, to the safety of whom works in the construction sites. Therefore, the designer shall extend his intervention so to define the correct anti-accident procedures of the works.

Chapter 2
Force Method

Abstract Starting from the basic laws of beam behaviour, the calculation methods of their flexural, torsional and axial deformations are presented and applied for the definition of the pertinent flexibility parameters. Using these parameters, the compatibility conditions are written, following the procedure of the Force Method, for the solution of hyperstatic simple beams, continuous beams and frame structures. Examples of the numerical solution of some type structures are eventually presented.

2.1 Flexural Deformations of Beams

Some indications are stated first about the reference systems and sign conventions that are adopted in the following. As indicated in Fig. 2.1, to a rectilinear beam, a *local system* of right-hand orthogonal axes is associated. The x-axis coincides with the central axis; the axes y and z coincide with the inertia principal axes of the cross section, so to locate the two planes (xy and xz) of right flexure. The origin is placed in the centre of the initial section "1" of the beam. The position of the current section is measured by the abscissa x. The end section "2" is located at the opposite extremity of the beam at a distance l.

The displacement of the current section j is defined by the three translations (measured in m):

δ_{xj}
δ_{yj}
δ_{zj}

and by the three rotations (measured in *radians*):

φ_x
φ_y
φ_z

the first three assumed positive if directed as the respective axes, the second three assumed positive if counter-clockwise for whom watches in the direction of the respective axes.

© Springer Nature Switzerland AG 2019
G. Toniolo, *Introduction to Frame Analysis*, Springer Tracts
in Civil Engineering, https://doi.org/10.1007/978-3-030-14664-1_2

The correspondent static quantities, distributed along the beam axis, represent the three forces (measured in N/m):

r_{xj}
r_{yj}
r_{zj}

and the three moments (measured in Nm/m):

m_{xj}
m_{yj}
m_{zj}

assumed positive following the same conventions of the geometric quantities.

To be noted that, in some literature, for rotations and moments, contrary conventions are assumed. To these latter, for example, all the most common procedures for automated structural analysis conform.

The deformation curves of the beam axis (see again Fig. 2.1) are represented by the functions (with $x_j = x$):

$$u(x) = \delta_{xj}$$
$$v(x) = \delta_{yj}$$

Fig. 2.1 Local reference system for a rectilinear beam

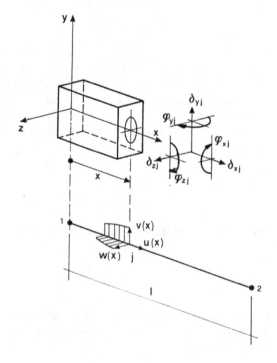

$$w(x) = \delta_{zj}$$

The first one is correlated to the *axial behaviour* of the beam; the others are correlated to its *flexural behaviour* in the two planes xy and xz. The *torsional behaviour* is correlated to the function

$$\varphi(x) = \varphi_{xj}$$

that represents the rotation of the sections around the x-axis of the beam.

For small displacements, the rotations in the two planes xy and xz of right flexure are obtained from:

$$\varphi_{yj} = +\frac{dw}{dx}$$
$$\varphi_{zj} = -\frac{dv}{dx}$$

For the deformations (measured in m/m the first three, in 1/m the second three)

ε_x (axial strain)
γ_y (shear strain along y)
γ_z (shear strain along z)
χ_x (torsional curvature)
χ_y (flexural curvature in xz)
χ_z (flexural curvature in xy)

and for the correspondent internal forces (measured in N the first three, in Nm the second three)

N   (axial force)
V_y (shear in xy)
V_z (shear in xz)
T   (torsional moment)
M_y (bending moment in xz)
M_z (bending moment in xy)

the same sign conventions of translations and rotations are valid if referred to the face that watches in the positive direction of the x-axis (Fig. 2.2).

In particular, one has:

$$\varepsilon_x = +\frac{du}{dx} \quad \text{(positive if of elongation)}$$

$$\chi_x = +\frac{d\varphi_x}{dx} \quad \text{(positive if of counter-clockwise spiral)}$$

and, for small displacements:

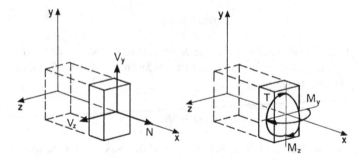

Fig. 2.2 Local reference system for internal forces

$$\chi_y = +\frac{d\varphi_y}{dx} = +\frac{d^2w}{dx^2}$$

$$\chi_z = +\frac{d\varphi_z}{dx} = -\frac{d^2v}{dx^2}$$

To be noted that, in the following, for the translation components of the joint displacements, sometime the symbols below are used:

ξ_i

η_i

ζ_i

considering them measured on the three axes *X, Y, Z* of a *global system* used as a reference of the overall structure, on the basis of specific definitions specified each time.

2.1.1 Short Notes on Mohr's Method

Reference is made to the simple case of Fig. 2.3. Only the flexural behaviour is to be analysed. The bearing supports represented in the figure correspond indifferently to hinges or trolleys since their horizontal effectiveness does not come into play.

Fig. 2.3 Representation of a general simply supported beam

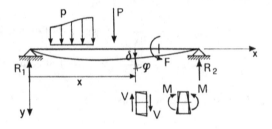

The y-axis is oriented downwards in the direction of the gravity loads, and, dealing with a problem contained in the xy plane, the symbols are simplified with respect to the three-dimensional case considered above; that is, it is set

$\delta = \delta_y$ (positive if directed downwards)
$\varphi = \varphi_z$ (positive if counterclockwise)
$p = r_y$ (uniformly distributed load positive if acting downwards)
P (point load positive if acting downwards)
F (point couple positive if counterclockwise)
$\gamma = \gamma_y$ (positive if clockwise)
$\chi = \chi_z$ (positive if with convexity at lower side)
$V = V_y$ (positive if clockwise)
$M = M_z$ (positive if the lower fibres are tensioned).

It is again set

R_1 $(=+V_1)$
R_2 $(=-V_2)$

support reactions positive if directed upwards, as those produced by the gravity loads.

It is reminded that, following the Elastic Theory of Beams, the deformation of an elementary beam segment long dx subjected to bending moment is given by

$$d\varphi = \frac{M}{EI}dx$$

where E is the longitudinal elastic modulus of the material and I $(=I_z)$ is the second moment of area of the section. So the curvature of the deformed axis, taking into account the bending moment only, is given by

$$\chi = \frac{M}{EI}$$

and varies along the beam consistently with the diagram of the bending moment.

For the calculation of the displacement components δ, and φ of the beam sections, the cited flexural curvatures are to be properly integrated. Let us consider, for example, the scheme of Fig. 2.4 on which the end rotations φ_1 and φ_2 are to be calculated, besides the deflection δ of the section of abscissa x. Due to the bending moment M $= M(\xi)$ of the current abscissa ξ, one has the deformation

$$d\varphi = \chi d\xi = \frac{M}{EI}d\xi$$

of the elementary segment long $d\xi$, and this deformation, in respect of the end supports of the beam, divides into parts in inverse proportions of the respective distances:

Fig. 2.4 Contribution of the
elementary curvature

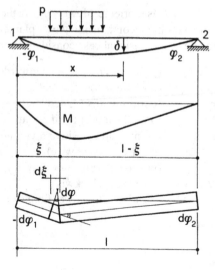

Fig. 2.5 Contribution of the
diffused curvature

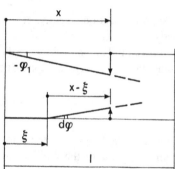

$$-d\varphi_1 = \frac{1-\xi}{1}d\varphi$$

$$+d\varphi_2 = \frac{\xi}{1}d\varphi$$

as can be easily verified in Fig. 2.4. In order to evaluate the total rotations, all the
elementary contributions have to be "summed up" with

$$-\varphi_1 = \int_0^1 \frac{1-\xi}{1}\chi\,d\xi$$

$$+\varphi_2 = \int_0^1 \frac{\xi}{1}\chi\,d\xi$$

The deflexion δ can be subsequently evaluated summing up the contributions of
the end rotation φ_1 and that of the diffused $d\varphi$ as indicated in Fig. 2.5:

$$\delta = -\varphi_1 x - \int_0^x (x - \xi)\chi d\xi$$

One can notice that the equations for the calculation of the end rotations written above are formally the same of those for the calculation of the support reactions due to a distributed load. From this observation it follows by consequence that, if the *curvature diagram* of the given beam is interpreted as an *equivalent load*

$$p^*(\xi) = \chi(\xi)$$

the consequent *equivalent reactions*

$$R_1^* = \int_0^l \frac{1 - \xi}{l} p^* d\xi$$

$$R_2^* = \int_0^l \frac{\xi}{l} p^* d\xi$$

are equal, except for the sign, to the actual rotations of the given beam (Fig. 2.6a).

$$-\varphi_1 = R_1^*$$
$$+\varphi_2 = R_2^*$$

One can notice again that the equation for the calculation of the deflection is formally the same of that for the calculation of the bending moment, and by consequence, one has (Fig. 2.6b):

Fig. 2.6 Effects of the equivalent load in terms of **a** end reactions and **b** bending moment

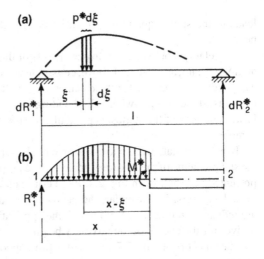

$$M^* = R_1^* x - \int_0^x (x - \xi) p^* d\xi = \delta$$

With a more general statement, one can say that, because of the formal equality of the compatibility relation between displacements and deformations

$$\frac{d^2 v}{dx^2} = -\chi$$

and the equilibrium relation between loads and internal forces

$$\frac{d^2 M}{dx^2} = -p,$$

if the curvature is interpreted as an equivalent load p*, the integration of the geometric relation necessary for the calculation of the displacements of the given beam can be replaced by the calculation of the internal forces of the *auxiliary beam* under the equivalent load, being

$$\delta = M^*$$
$$\varphi = -V^*$$

In particular, the second relation, written for the end sections of the beam on which one has

$$V_1^* = +R_1^*$$
$$V_2^* = -R_2^*$$

leads to the same equations written before for the support reactions of the given beam.

The relations between the displacements of the given beam and the forces of the auxiliary beam, as presented above, constitute the *Mohr's theorem* and its *corollaries*. They indicate a method for the calculation of deflections and rotations that, applied to the isostatic beams, where the bending moment diagram is definable with the basic equations of the statics, is particularly easy, as it will result from the following examples.

To be noted that, in order to conform with the support conditions of the given beam, the static scheme of the auxiliary beam is to be properly adapted following the criteria presented here below. In Fig. 2.7, these criteria are described for the different types of end supports. In particular, the degrees of flexural effectiveness of the supports are indicated in brackets, from which the applicative rule of the "complement to 2" derives for the choice of the auxiliary beam.

The case (a) of Fig. 2.7 refers to a full support that prevents both the rotation and the translation and becomes a free end in the auxiliary beam so that the correspondent equivalent reactions keep null; in the case (b), the support in the auxiliary

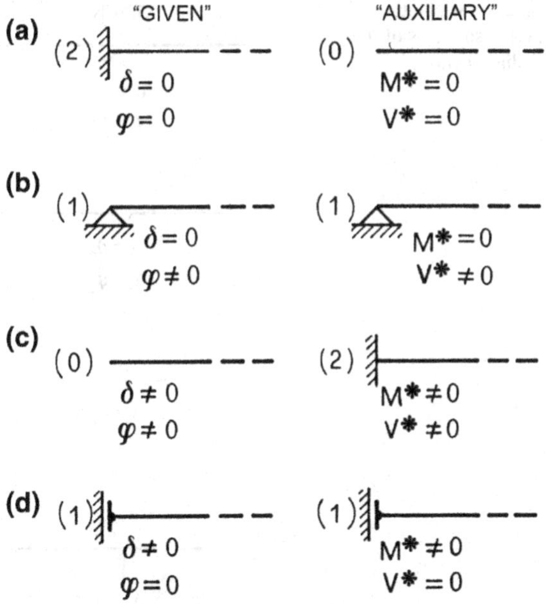

Fig. 2.7 Adaptation of the end supports of the auxiliary beam

$$\textbf{(a)} \quad (2) \quad \overset{\text{"GIVEN"}}{\underset{\substack{\delta = 0 \\ \varphi = 0}}{\rule{0pt}{0pt}}} \qquad (0) \quad \overset{\text{"AUXILIARY"}}{\underset{\substack{M^* = 0 \\ V^* = 0}}{\rule{0pt}{0pt}}}$$

$$\textbf{(b)} \quad (1) \underset{\substack{\delta = 0 \\ \varphi \neq 0}}{\rule{0pt}{0pt}} \qquad (1) \underset{\substack{M^* = 0 \\ V^* \neq 0}}{\rule{0pt}{0pt}}$$

$$\textbf{(c)} \quad (0) \underset{\substack{\delta \neq 0 \\ \varphi \neq 0}}{\rule{0pt}{0pt}} \qquad (2) \underset{\substack{M^* \neq 0 \\ V^* \neq 0}}{\rule{0pt}{0pt}}$$

$$\textbf{(d)} \quad (1) \underset{\substack{\delta \neq 0 \\ \varphi = 0}}{\rule{0pt}{0pt}} \qquad (1) \underset{\substack{M^* \neq 0 \\ V^* = 0}}{\rule{0pt}{0pt}}$$

beam remains unchanged so to have only the equivalent reaction (shear) force that corresponds to the rotation of the given beam; the case (c) represents the obvious dual situation of case (a), while the sliding support of the case (d) remains unchanged so to have only the equivalent moment component that corresponds with the translation of the given beam.

Figure 2.8 describes some possible cases of internal supports; the similar adaptation criteria of the auxiliary beam are clearly indicated by the captions (the subscripts "s" and "d" mean "left" and "right"), without the need of further explanations,

The simple example of Fig. 2.9 is finally proposed with reference to a cantilever beam, long l, with constant cross section (EI = const.) loaded with a couple F at its free end.

For this "given" beam, the diagram of the bending moment is constant:

$$M = -F \quad \text{(the upper fibres are tensioned)}$$

In the auxiliary beam, the end supports are inverted following the rule given above; on this beam, the equivalent load is applied equal to

$$p^* = \frac{M}{EI} = \frac{-F}{EI} \quad \text{(with upwards direction)}$$

Fig. 2.8 Adaptation of the
internal supports of the
auxiliary beam

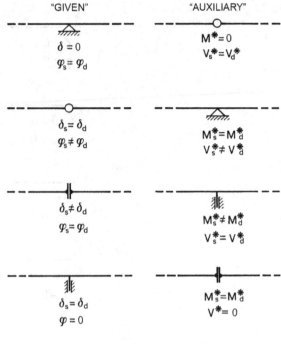

"GIVEN"

$\delta = 0$
$\varphi_s = \varphi_d$

$\delta_s = \delta_d$
$\varphi_s \neq \varphi_d$

$\delta_s \neq \delta_d$
$\varphi_s = \varphi_d$

$\delta_s = \delta_d$
$\varphi = 0$

"AUXILIARY"

$M^* = 0$
$V_s^* = V_d^*$

$M_s^* = M_d^*$
$V_s^* \neq V_d^*$

$M_s^* \neq M_d^*$
$V_s^* = V_d^*$

$M_s^* = M_d^*$
$V^* = 0$

Fig. 2.9 Example of a
cantilever beam

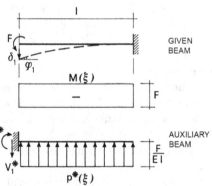

At the fixed end, the equivalent reaction are then evaluated:

$$M_1^* = -\frac{p^* l^2}{2} = +\frac{Fl^2}{2EI}$$

$$V_1^* = +p^* l = -\frac{Fl}{EI} = R_1^*$$

Finally, for the end section "1" of the given beam, one obtains:

$$\delta_1 = +M_1^* = +\frac{Fl^2}{2EI} \quad \text{(with downwards direction)}$$

$$\varphi_1 = -V_1^* = +\frac{Fl}{EI} \quad \text{(counter-clockwise)}$$

2.1.2 Calculation of Beam Flexibilities

Another example of application of the method is shown with reference to the simply supported beam of Fig. 2.10. For the action of the given couple F at the end 2, one has the linear diagram of bending moment indicated in the figure.

On the auxiliary beam, that keeps in this case the same end supports as the given one, the equivalent load is applied with a triangular distribution; its maximum intensity is

$$p_o^* = \frac{F}{EI}$$

The load resultant is

$$Q^* = p_o^* \frac{1}{2} = \frac{Fl}{2EI}$$

and it is placed at $l/3$ from the right support. The corresponding reactions are immediately obtained:

Fig. 2.10 Example of a simply supported beam

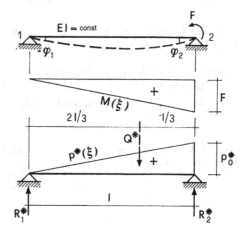

$$R_1^* = \frac{1/3}{1}Q^* = \frac{Fl}{6EI}$$

$$R_1^* = \frac{21/3}{1}Q^* = \frac{Fl}{3EI}$$

that, on the basis of the corollaries of Mohr's theorem, are equal, except the sign, to the end rotations of the given beam:

$$-\varphi_1 = R_1^* = \frac{1}{6EI}F$$

$$\varphi_2 = R_2^* = \frac{1}{3EI}F$$

The simply supported arrangement is analysed with these procedures in order to provide some characteristic parameters of the flexural behaviour of the beams: first the *flexibilities* of the end sections, defined as end rotations due to couples of unit value applied to the same sections.

In Fig. 2.11, these quantities are described in details. The sign conventions assume positive the flexibilities d consistently with the downwards deflection indicated in the figure, so that at the first end the sign of the correspondent rotation is changed:

$$d_1 = -\varphi_1$$

$$d_2 = +\varphi_2$$

In the same way, the unit flexural actions are assumed, so that they are positive if the lower fibres are tensioned as for the bending moments:

$$M_1 = -F_1$$

$$M_2 = +F_2$$

So the flexural behaviour of a beam is represented by the three parameters described in Fig. 2.11:

– the *direct flexibility* of the first end d_1;
– the *indirect flexibility* d_i;
– the *direct flexibility* of the second end d_2.

Fig. 2.11 Representation of the flexibilities of the beam

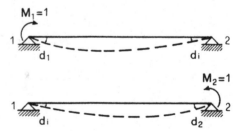

The two direct flexibilities, that express the rotation on the end of application of the unit couple, are mutually equal in the case of symmetrical beams. The indirect flexibilities are in any case "unique", as demonstrated by the pertinent "reciprocity" theorems. In particular, for beams with constant section, as deducible from the example shown above setting $F = 1$, one has

$$d_1 = d_2 = \frac{1}{3EI}$$

$$d_i = \frac{1}{6EI}$$

For beams with variable section, proper numerical procedures have to be applied for the curvature integration, as already shown in Sect. 2.1.1, setting, respectively, the expressions (see Fig. 2.12):

$$\chi(\xi) = \frac{1 - \xi/l}{EI(\xi)}$$

$$\chi(\xi) = \frac{\xi/l}{EI(\xi)}$$

and discretizing the problem with a subdivision of the beam in n segments with length $\Delta\xi = l/n$. So, for example, one can set:

$$d_1 = \int_0^l \left(\frac{1-\xi}{l}\right)^2 \frac{d\xi}{EI(\xi)} \simeq \sum_{i=1}^n \left(\frac{1-\xi_i}{l}\right)^2 \frac{\Delta\xi}{EI_i}$$

$$d_1 = \int_0^l \frac{\xi}{l}\frac{1-\xi}{l} \frac{d\xi}{EI(\xi)} \simeq \sum_{i=1}^n \frac{\xi_i}{l}\frac{1-\xi_i}{l} \frac{\Delta\xi}{EI_i}$$

Fig. 2.12 Discretized beam
of variable section

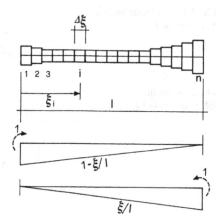

$$d_1 = \int_0^1 \left(\frac{\xi}{1}\right)^2 \frac{d\xi}{EI(\xi)} \simeq \sum_{i=1}^n \left(\frac{\xi_i}{1}\right)^2 \frac{\Delta\xi}{EI_i}$$

Besides the flexibilities, the *end rotations* are analysed; they just represent the rotations $\overline{\varphi}_1$ and $\overline{\varphi}_2$ at the ends of the simply supported beam due to the different types of load (see Fig. 2.13). With the same sign conventions that assume positive the rotation of the first end if clockwise and of the second end if counterclockwise, one can set

$$\overline{\varphi}_1 = \int_0^1 \frac{1-\xi}{1} \frac{M(\xi)}{EI(\xi)} d\xi$$

$$\overline{\varphi}_2 = \int_0^1 \frac{\xi}{1} \frac{M(\xi)}{EI(\xi)} d\xi$$

and, for the bending moment $M(\xi)$ (Fig. 2.14), the first end reaction is previously evaluated with:

$$R_1 = \sum_i \frac{F_i}{1} + \sum_j \frac{1-b_j}{1} P_j + \sum_k \frac{1-c_k/2}{1} e_k p_k$$

and then the following expression can be elaborated

$$M(\xi) = +R_1\xi - \sum_i F_i U(\xi - a_i) - \sum_j P_j(\xi - b_j)U(\xi - b_j)$$

$$- \sum_k p_k \frac{(\xi - c_k)^2}{2} U(\xi - c_k)$$

Fig. 2.13 End rotations for a general load

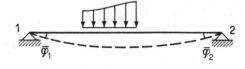

Fig. 2.14 Beam with different types of load

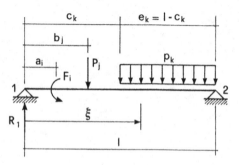

Fig. 2.15 Case of symmetric beam with point load

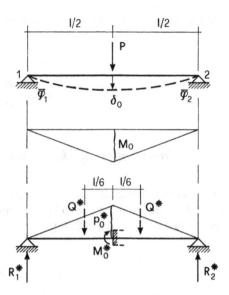

where $U(x)$ expresses the "step function" that is 0 for $x < 0$ and 1 for $x > 0$.

Section 2.4 shows the analysis of different elementary cases of beams with constant cross section subjected to the most common types of load. To conclude this, preliminary presentation only one case of a symmetric beam with a point load placed at mid-span is shown (see Fig. 2.15).

To the bilinear diagram of bending moment, with a maximum value

$$M_o = \frac{Pl}{4}$$

corresponds on the auxiliary beam a similar bilinear scheme of equivalent load with a maximum value

$$P_o^* = \frac{Pl/4}{EI}$$

The semi-resultant of the load is

$$Q^* = \frac{1}{2}P_o^* l/2 = \frac{Pl^2}{16EI}$$

and corresponds, due to symmetry, to the support reactions $(R_1^* = R_2^* = Q^*)$. Eventually, the bending moment at mid-span is calculated with

$$M_o^* = R_1^* l/2 - Q^* l/6 = \frac{Pl^3}{48EI}$$

So the end rotations and the mid-span deflection are

$$\overline{\varphi}_1 = \overline{\varphi}_2 = \frac{Pl^2}{16EI}$$

$$\delta_o = \frac{Pl^3}{48EI}$$

2.1.3 Contribution of Shear Deformation

The displacement calculations presented in the preceding sections took into account only of the deformation due to bending moment. The deformation due to shear is represented in Fig. 2.16a and, within an elastic behaviour of the sections, is given by

$$dv = \frac{V}{GA^*}d\xi$$

where G is the tangent elastic modulus and A^* $(=A_y)$ is the "shear area", which is obtained from the cross section area A reduced by the shear factor χ $(=\chi_y)$ that depends on the shape of the section:

$$A^* = \frac{A}{\chi}$$

Fig. 2.16 Shear deformation of a beam

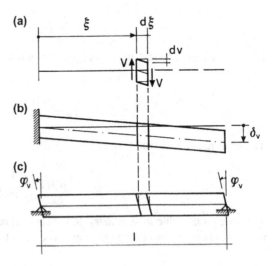

In order to calculate the overall deformation (e.g., the translation δ_v of Fig. 2.16b), one has to integrate the elementary shear deformation

$$\gamma = \frac{V}{GA^*}$$

over the length of the beam

$$\delta_v = \int_0^1 \gamma d\xi$$

If one has, instead a cantilever beam, a simply supported beam (see Fig. 2.16c), for the compatibility of the supports, a rotation

$$\varphi_v = \frac{\delta_v}{1}$$

should be added the same for all the sections.

Let's consider the example of the cantilever beam with constant section described in Fig. 2.17; at the left side, the deformation due to bending moment is shown; at the right side, the deformation due to shear is shown. The former are evaluated, following the Mohr's method presented before, on the basis of the bending moment diagram which maximum value is

$$M_o = -Pl$$

On the auxiliary beam, that has the end supports interchanged, the triangular distribution of load is applied in the upwards direction with a maximum value

$$p_o^* = -\frac{Pl}{EI}$$

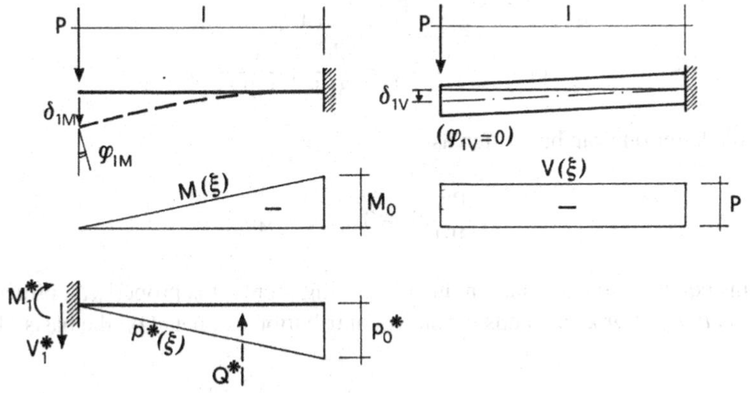

Fig. 2.17 Flexural and shear deformation contributions

The resultant of the equivalent load is

$$Q^* = \frac{1}{2}p_o^* l = -\frac{Pl^2}{2EI}$$

and it is placed at a distance $2l/3$ from the left full support. So on this support, one has the components

$$V_1^* = Q^*$$
$$M_1^* = -Q^* 2l/3$$

that give the displacement contributions

$$\varphi_{1M} = -V_1^* = \frac{pl^2}{2EI} \quad \text{(counter-clockwise)}$$

$$\delta_{1M} = M_1^* = \frac{Pl^3}{3EI} \quad \text{(in downwards direction)}$$

In order to evaluate the shear contributions, the correspondent diagram is drawn down (see again Fig. 2.17):

$$V(\xi) = -P \quad (= \text{const.})$$

obtaining by consequence

$$\delta_{1V} = \int_1^0 \frac{V(\xi)}{GA^*} d\xi = \int_0^1 \frac{P}{GA^*} d\xi = \frac{Pl}{GA^*}$$

with $\varphi_{1V} = 0$.

So globally, one has

$$\varphi_1 = \varphi_{1M} + \varphi_{1V} = \frac{Pl^2}{2EI}$$

$$\delta_1 = \delta_{1M} + \delta_{1V} = \frac{Pl^3}{3EI} + \frac{Pl}{GA^*}$$

where the latter one can be written as

$$\delta_1 = \frac{Pl^3}{3EI}(1 + 3\kappa_o) = \delta_{1M}\kappa$$

In this equation, the translation due to bending moment is properly amplified by the *correction factor* κ that adds the shear contribution evaluated on the basis of the ratio

Fig. 2.18 Shear contribution to deformation

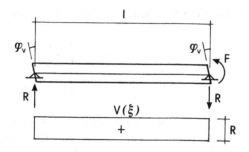

$$\kappa_o = \frac{EI}{l^2 GA^*}$$

between the elastic flexural and shear deformation properties of the beam.

In a similar way, the end rotations of the beam of Fig. 2.10 can be corrected. For this beam in Fig. 2.18, the shear deformation contribution is shown; it should be added to the flexural contribution already defined in Sect. 2.1.2.

For the action of the couple F, one has at the end supports two equal and opposite reactions

$$R = \frac{F}{l}$$

that provide the constant shear force

$$V(\xi) = R$$

So setting

$$\delta_V = \int_0^l \frac{V(\xi)}{GA^*} d\xi = \frac{F}{GA^*}$$

one obtains the rotation

$$\varphi_V = \frac{\delta_V}{l} = \frac{F}{lGA^*}$$

that has to be added to the flexural contribution.

With the sign conventions of the flexibilities, one has

$$\overline{\varphi}_1 = \frac{Fl}{6EI} - \frac{F}{lGA^*} = \frac{Fl}{6EI}(1 - 6\kappa_o)$$

$$\overline{\varphi}_2 = \frac{Fl}{3EI} + \frac{F}{lGA^*} = \frac{Fl}{3EI}(1 + 3\kappa_o)$$

expressions that, with $F = 1$, give also the *corrected flexibilities*

$$d_i = \frac{1}{6EI}2\left(\frac{1}{2} - 3\kappa_o\right) = \frac{1}{3EI}\kappa_i$$

$$d_2 = \frac{1}{3EI}(1 + 3\kappa_o) = \frac{1}{3EI}\kappa$$

where one can see that the shear deformation contribution increases the direct flexibility and decreases in the same time the indirect flexibility.

In order to show the magnitude order of the correction, the ratio κ_o is evaluated as a function of the slenderness $\lambda = l/h$ of the beam, having indicated with h the depth of its cross section. With a Poisson factor ν varying for the different materials from 0.15 to 0.30, one obtains in average

$$\frac{E}{G} = 2(1 + \nu) = 2.3 \text{ to } 2.6 \approx 5/2$$

First, a section shape "very stiff" for shear is examined, like the one of Fig. 2.19a. In this case, one has, with $A^* = 5A/6$:

$$\frac{I}{A^*} = \frac{Ah^2/12}{5A/6} = \frac{h^2}{10}$$

obtaining by consequence

$$\kappa_o = \frac{5}{2}\frac{h^2}{10}\frac{1}{l^2} = \frac{1}{4}\frac{1}{\lambda^2}$$

This value leads, for the correction factors κ and κ_i, to the curves of Fig. 2.20.

It can be noted that these corrections are negligible for medium-high slenderness, they are instead important in the field of the deep beams. It can be noted again that the indirect slenderness, when corrected with the pertinent factor κ_i, decreases with the lower slenderness so to change sign when the deformation effect of shear becomes larger than the one of bending moment.

For a section "little stiff" for shear, as the one with I shape of Fig. 2.19b, assuming, with $c \ll h$ and $t \ll b$,

Fig. 2.19 Different shapes of cross sections

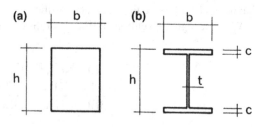

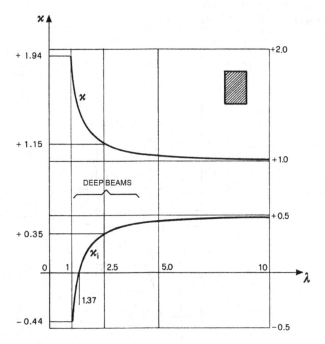

Fig. 2.20 Curves of correction factors for very stiff sections

$$A' \approx 2\,bc \approx 0.67\,A$$
$$A^* \approx h\,t \approx 0.33\,A$$
$$I \approx 2\,bc\,h^2/4 \approx 0.67\,A\,h^2/4,$$

one obtains

$$\frac{I}{A^*} \approx \frac{0.67Ah^2/4}{0.33A} = \frac{h^2}{2}$$

and by consequence

$$\kappa_0 = \frac{5}{2}\frac{h^2}{2}\frac{1}{l^2} = \frac{5}{4}\frac{1}{\lambda^2}$$

For the correction factors, one has this time the curves of Fig. 2.21, where the larger extension can be noted of the deep beams for which the deformation influence of shear is relevant.

Fig. 2.21 Curves of
correction factors for little
stiff sections

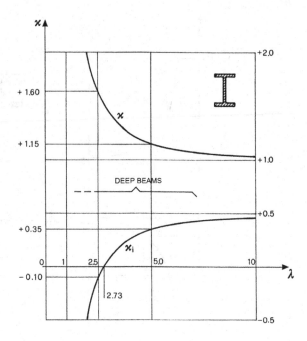

2.2   Compatibility Equations

For an isostatic beam, the calculation of the support reactions and internal forces due
to a given load distribution can be performed with the sole fundamental equations of
the statics; once the internal forces are known, for the calculation of the displacement,
the equation of the elastic line can be set, for example, in the form

$$\frac{d^2v}{dx^2} = -\frac{M(x)}{EI}$$

valid for beams slender enough so that the shear deformation can be neglected. Of
this equation in the preceding section, a handy solution method has been shown based
on Mohr's theorem.

For hyperstatic beams, like the one of Fig. 2.22, the internal forces become dependent from the deformation behaviour; so, being not known the bending moment
function $M(x)$, the equation of the elastic line should be set directly on the applied
loads, for example, in the form

$$\frac{d^4v}{dx^4} = \frac{p(x)}{EI}$$

valid for beams with constant section.

The integration of this differential equation does not present relevant difficulties. But, for the definition of the integration constants, it leads to more onerous calculations with respect to other solution methods.

One could move, for example, the problem into the easier field of the isostatic beams, reserving the application of the proper additional conditions of equivalence with the given hyperstatic problem. In this way, with reference to the beam of Fig. 2.23, the redundant end restrains can be deleted moving to the simply supported arrangement of Fig. 2.23a.

The equivalence of this auxiliary beam with the given one can be set if on the former beam the static actions suppressed together with the redundant restrains are expressly applied as external loads. In this way, referring again to Fig. 2.23a, in addition to the given load $p(x)$ the two bending moments X_1 and X_2 that the original end supports could transmit are applied.

These moments are the *static unknowns* of the problem following the method that is here presented. They are applied in a self-balanced way that fulfils the equilibrium conditions at the end supports. Among all their possible balanced values, the only ones that fulfil also the other equivalence condition shall be defined. So the geometric

Fig. 2.22 Hyperstatic beam

Fig. 2.23 Equivalence conditions

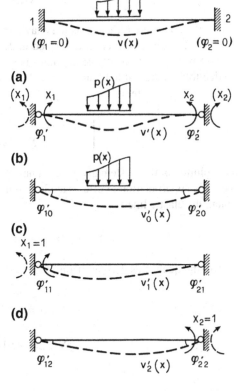

compatibility is to be imposed to the auxiliary beam at the end sections where it is different from the given beam. As in the given beam, one had (see Fig. 2.49):

$$\varphi_1 = 0$$
$$\varphi_2 = 0$$

so for the auxiliary beam, it shall be set (see Fig. 2.23a).

$$\varphi_1' = 0$$
$$\varphi_2' = 0$$

With these conditions, the equivalence is fully fulfilled and the elastic lines of both the beams, the given hyperstatic one and the auxiliary isostatic one, become the same:

$$v(x) = v'(x)$$

In order to lay down the *compatibility conditions*

$$\begin{cases} \varphi_1'(p; X_1, X_2) = 0 \\ \varphi_2'(p; X_1, X_2) = 0 \end{cases},$$

within the considered linear elastic behaviour, the superposition and the proportionality of effects can be used as described in details in Fig. 2.23b–d. That is, for each of the two end sections of the auxiliary beam, the rotation contribution is summed up due to the external load and the single static unknowns, obtaining the following linear equation system

$$\begin{cases} \varphi_{10}' + \varphi_{11}'X_1 + \varphi_{12}'X_2 = 0 \\ \varphi_{20}' + \varphi_{21}'X_1 + \varphi_{22}'X_2 = 0 \end{cases}$$

The solution of this system gives the values of the unknowns, and consequently, with the same superposition of effects applied to the isostatic auxiliary structure, one can obtain the elastic line

$$v(x) = v_{10}'(x) + v_{11}'(x)X_1 + v_{12}'(x)X_2$$

and all the quantities deriving from it:

$$M(x) = -EI\frac{d^2v}{dx^2}$$
$$V(x) = -EI\frac{d^3v}{dx^3}$$

With this introduction of the *Force Method,* its operative procedure is presented as applicable to the practical analysis of the hyperstatic beams. To be noted that, giving for granted the equivalence problem presented here above, the apex given to the quantities of the auxiliary beam will no more used.

2.2.1 Hyperstatic Beams

The first applications of the Force Method are presented with reference to the simple cases of one span hyperstatic beams. So let's consider the beam of Fig. 2.24 provided with a full restraint at one end and a simple support at the other end. This support arrangement has one degree of redundancy. The beam has a constant cross section (EI = const.), and a clockwise couple F is applied to its second end.

The isostatic auxiliary beam is obtained deleting the rotation restraint and applying the correspondent hyperstatic moment X_1 (see Fig. 2.25a). On this beam, the superposition of effects is performed, applying first the given load F (see Fig. 2.25b) and then the unitary static unknown $X_1 = 1$ (see Fig. 2.25c). For the compatibility with the given beam, one shall impose a zero value for the rotation at the end 1 where the rotation restraint has been deleted:

$$\varphi_{11} X_1 + \varphi_{10} = 0$$

Fig. 2.24 Beam with a full restraint and a simple support

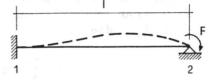

Fig. 2.25 Superimposition of effects on the auxiliary beam

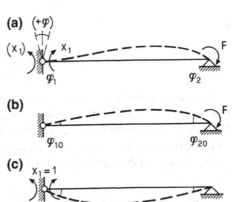

For the sign conventions, the hyperstatic moment is assumed positive if it tensions the lower fibres, and the corresponding rotation between the beam end and the adjacent external restraint is assumed positive if concordant with the moment, that is, if the rotation φ is opened at the upper side (see Fig. 2.25a).

The coefficients of the compatibility equation, as described in Fig. 2.25b, c, are taken from the example of Sect. 2.1.2 (see Fig. 2.10) and are

$$\varphi_{11} = \frac{1}{3EI} \quad \text{(direct flexibility)}$$

$$\varphi_{10} = \frac{1}{6EI}F \quad \text{(rotation due to load)}$$

The solution is obtained with

$$X_1 = -\frac{\varphi_{10}}{\varphi_{11}} = +\frac{F}{2}$$

and shows that, for the action of a clockwise couple F applied to the simply supported end of the beam, to the opposite built-in end, a clockwise halved moment X_1 is transmitted. The value

$$t_{12} = \frac{d_i}{d_2} = 0.5$$

of the moment *transmission factor* is valid for straight beams with constant cross section and with a sufficient slenderness so that the deformation shear contribution is negligible. For squat beams, the "corrected" value, as deduced from what presented at Sect. 2.1.3, would be

$$t_{12} = \frac{\kappa_i}{\kappa} = \frac{1}{2}\frac{1 - 6\kappa_0}{1 + 3\kappa_0} < 0.5$$

with

$$\kappa_0 = \frac{EI}{l^2 GA^*}$$

Remaining in the field of the slender beams, the solution is completed with the diagrams of Fig. 2.26: For the bending moment, one has a linear diagram, drawn at the tension side, starting from the "positive" value $M_1 = +F/2$, that becomes null at $1/3$ of the span and that arrives to the "negative" value $M_2 = -F$. For the shear force, one has a constant diagram with a "negative" value (counterclockwise) equal to

$$V = -R = -\frac{1}{l}\left(\frac{F}{2} + F\right) = -\frac{3}{2}\frac{F}{l}$$

Fig. 2.26 Bending moment and shear diagrams

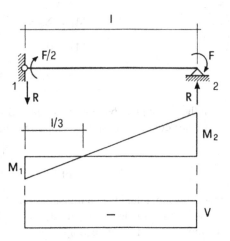

At last, one can calculate the rotation at the simply supported end of the beam superposing the effects of the load and the hyperstatic:

$$\varphi_2 = -\frac{1}{3EI}F + \frac{1}{6EI}X_1 = -\frac{1}{4EI}F$$

This rotation is clockwise like the applied couple F. In particular, setting $F = 1$, the rotation

$$d = \frac{1}{4EI}$$

can be assumed as the *direct flexibility* of the simply supported end of the beam of concern. This flexibility is obviously lower than the one referred to the doubly supported beam analysed at Sect. 2.1.2 (see Fig. 2.10).

Let's consider now the same beam just examined subjected to the geometric actions represented in Fig. 2.27. In the solution procedure only the term φ_{10} due to the load shall be changed; the coefficient φ_{11} related to the structure remains the same.

Due to the applied support rotation ψ of Fig. 2.27a, one has, as described in Fig. 2.28, the relative rotation

$$\varphi_{10} = -\psi$$

negative following the sign conventions defined before (see Fig. 2.25a). By consequence, the solution is

$$X_1 = -\frac{\varphi_{10}}{\varphi_{11}} = +\frac{3EI}{l}\psi$$

$$M_1 = X_1 = +\frac{3EI}{1}\psi$$

$$M_2 = 0$$

$$V = -R = -\frac{1}{1}X_1 = -\frac{3EI}{1^2}\psi$$

with the consequent bending moment and shear diagrams shown in Fig. 2.28.
   In particular, at the simply supported end, one has the rotation

$$\varphi_2 = 0 + \frac{1}{6EI}X_1 = +\frac{\psi}{2} \quad \text{counter-clockwise}$$

that shows in a dual way a transmission of the rotation ψ from one end to the other
of the beam with the same factor $t_{12} = 0.5$ that governs the opposite transmission of
the moment F of the preceding case.
   Due to the linear displacement δ of the left support of Fig. 2.27b, in the auxiliary
beam, one has, as described in Fig. 2.29, a relative rotation

$$\varphi_{10} = -\frac{\delta}{1}$$

Fig. 2.27 Beam subjected to
geometric actions

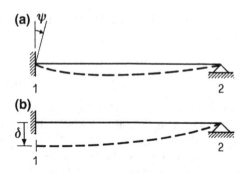

Fig. 2.28 Case of an applied
support rotation

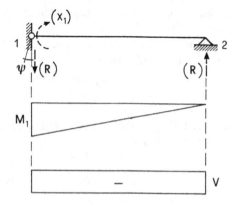

negative following the same sign conventions. By consequence, one obtains

$$X_1 = \frac{\varphi_{10}}{\varphi_{11}} = +\frac{3EI}{l^2}\delta$$

$$M_1 = X_1 = \frac{3EI}{l^2}\delta \quad M_2 = 0$$

$$V = -R = \frac{1}{l}X_1 = -\frac{3EI}{l^3}\delta$$

with the diagrams of bending moment and shear force drawn in the same Fig. 2.29. In particular, at the opposite end of the beam, one has the rotation

$$\varphi_2 = \frac{\delta}{l} + \frac{1}{6EI}X_1 = \frac{3}{2}\frac{\delta}{l} \quad \text{(counter-clockwise)}$$

evaluated with the same superposition of effects.

Let's consider now the beam with two full end supports under the three load conditions shown in Fig. 2.30b, c. The effects of the hyperstatics X_1 and X_2 on the auxiliary beam are shown in Fig. 2.31a, b. One has the following system of the two compatibility conditions

$$\begin{cases} \frac{1}{3EI}X_1 + \frac{1}{6EI}X_2 + \varphi_{10} = 0 \\ \frac{1}{6EI}X_1 + \frac{1}{3EI}X_2 + \varphi_{20} = 0 \end{cases}$$

The solution is

$$X_1 = -\frac{2EI}{l}(2\varphi_{10} - \varphi_{20})$$

Fig. 2.29 Case of an applied support linear displacement

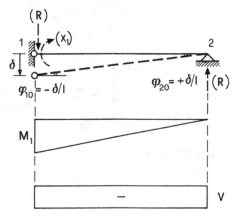

$$X_2 = -\frac{2EI}{1}(2\varphi_{20} - \varphi_{10})$$

For the three given load conditions, one has, respectively, in the auxiliary beam the relative rotations indicated in Fig. 2.32a–c, where in particular the ones due to the load concentrated at mid-span are deduced from the last example of Sect. 2.1.2 (see Fig. 2.15), and the others follow directly from the kinematics of Fig. 2.32b, c.

So for the first load condition, one has

$$\varphi_{10} = \varphi_{20} = \frac{Pl^2}{16EI};$$

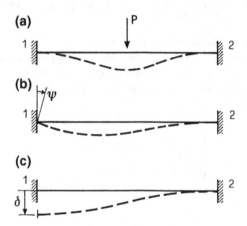

Fig. 2.30 Case of beam with two full end supports

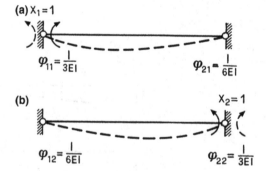

Fig. 2.31 Effects of the hyperstatics

Fig. 2.32 Relative rotations due to loads

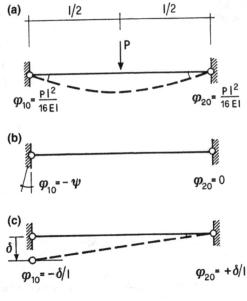

Fig. 2.33 Diagrams of internal forces

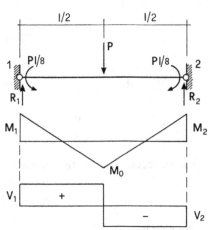

the solution is

$$X_1 = X_2 = -\frac{Pl^2}{8}$$

and leads to the diagrams of the internal forces drawn in Fig. 2.33, where one has

Fig. 2.34 Diagrams of
internal forces

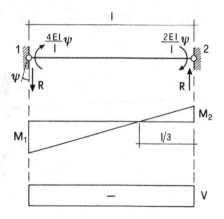

$$M_1 = X_1 = -\frac{Pl}{8} = M_2 \qquad \text{(tensioned fibres at upper side)}$$

$$M_o = \frac{Pl}{4} + \frac{X_1 + X_2}{2} = +\frac{Pl}{8} \quad \text{(tensioned fibres at the lower side)}$$

$$R_1 = R_2 = +\frac{P}{2} \qquad \text{(oriented in upwards direction)}$$

$$V_1 = R_1 = \frac{P}{2} \qquad \text{(clockwise)}$$

$$V_2 = -R_2 = -\frac{P}{2} \qquad \text{(counter-clockwise)}$$

For the second load condition, one has

$$\varphi_{10} = -\psi \quad \varphi_{20} = 0;$$

the solution is

$$X_1 = +\frac{4EI}{l}\psi$$
$$X_2 = -\frac{2EI}{l}\psi$$

and leads to the diagrams of the internal forces drawn in Fig. 2.34, where one has

$$M_1 = X_1 = +\frac{4EI}{1}\psi \qquad \text{(tensioned fibres at lower side)}$$

$$M_2 = X_2 = -\frac{2EI}{1}\psi \quad (= -M_1/2) \quad \text{(tensioned fibres at upper side)}$$

$$V = -R_1 = -\frac{X_1 - X_2}{1} = -\frac{6EI}{1^2}\psi \quad \text{(counter-clockwise)}$$

In particular, one can see, between the end of application of the rotation ψ and the opposite end of the beam, the same moment transmission found (in the opposite direction) for the case of Fig. 2.26.

For the third load condition, one has

$$\varphi_{10} = -\varphi_{20} = -\frac{\delta}{1}$$

and the solution is

$$X_1 = +\frac{6EI}{1^2}\delta$$

$$X_2 = -\frac{6EI}{1^2}\delta$$

leading to the diagrams of internal forces drawn in Fig. 2.35, where one has

Fig. 2.35 Diagrams of internal forces

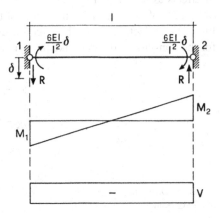

$$M_1 = X_1 = +\frac{6EI}{l^2}\delta \qquad\qquad \text{(tensioned fibres at lower side)}$$

$$M_2 = X_2 = -\frac{6EI}{l^2}\delta \;\; (= -M_1) \qquad \text{(tensioned fibres at upper side)}$$

$$V = -R = -\frac{X_1 - X_2}{l} = -\frac{12EI}{l^3} \;\;\text{(counter-clockwise)}$$

Other different examples are present in Sect. 2.4.1.

2.2.2   Standard Procedure for Continuous Beams

Obviously, the Force Method can be applied with different choices of the auxiliary structure. For example, in the beam with one redundancy of Fig. 2.24, one can delete the support at the second end instead of the rotation restraint at the first end of the beam. For such an auxiliary beam (see Fig. 2.36), one should impose the compatibility condition of the vertical displacement at the end 2:

$$\delta_2 = 0$$

So in order to express the relative displacement between the bearing and the adjacent beam end that has been disconnected, one applies again the superposition of effects as shown in Fig. 2.37. The values are taken, respectively, from the case of Fig. 2.9 of Sect. 2.1.1 for the effect of the applied couple F and from the case of Fig. 2.17 of Sect. 2.1.3 for the effect of the hyperstatic X, assuming a positive sign for the translations directed like the hyperstatic itself:

$$\delta_{20} = -\frac{Fl^2}{2EI}$$

$$d_{2X} = \frac{l^3}{3EI}$$

Fig. 2.36 Different auxiliary beam

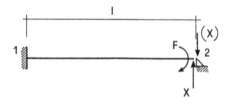

Fig. 2.37 Superimposition of effects

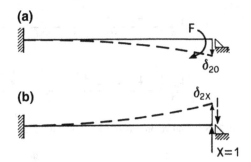

From the compatibility equation

$$\delta_{2X} X + \delta_{20} = 0$$

one obtains the solution

$$X = -\frac{\delta_{20}}{\delta_{2X}} = +\frac{3}{2}\frac{F}{l}$$

that, with its positive sign, confirms the direction evidenced in Fig. 2.36 and leads to the same values

$$M_1 = -F + Xl = +F/2$$
$$V = -X = -1.5\,F/l$$

obtained at Sect. 2.2.1 using the auxiliary beam of Fig. 2.25.

For the choice of the auxiliary structure, one should consider that the better efficiency of the Force Method, with respect to other methods, is based on the possibility of an immediate definition of the coefficients of the solving equations. And this can be obtained if a standard criterion is assumed for the deletion of the redundant restrains so that few recurrent expressions can be used taken from available tables.

So following this criterion, for the continuous beams such as that represented in Fig. 2.38, the *standard procedure* of the Force Method is based on the deletion of the restraints of flexural continuity over the external supports. In this way, the structure is reduced to an assembly of beams all with end hinges and mutually independent. For the formulation of the coefficients, one can refer to one only type of beam (the simply supported one), analysing any single span separately.

Figure 2.39 describes the analysis for the formulation of the compatibility equation at the kth support of the beam, based on the usual superposition of effects that expresses the relative rotation between the involved contiguous ends of the spans "c" and "e" as:

$$\varphi_k = \varphi_{kj} X_j + \varphi_{kk} X_k + \varphi_{kn} X_n + \varphi_{k0}$$

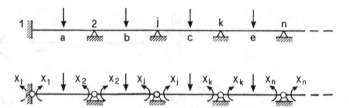

Fig. 2.38 Standard procedure for continuous beam

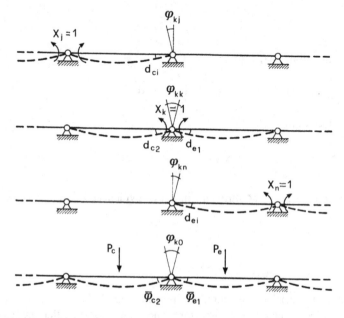

Fig. 2.39 Superimposition of effects on the kth support

With the notations of Fig. 2.39, the compatibility equation becomes

$$d_{ci}X_j + (d_{c2} + d_{e1})X_k + d_{ei}X_n + (\overline{\varphi}_{c2} + \overline{\varphi}_{e1}) = 0$$

For beams with constant cross section, one can see how all the coefficients of the unknowns are composed with the only two expressions of the flexibilities, direct $d_1 = d_2$ and indirect d_i deduced at Sect. 2.1.2:

$$d_{c2} = \frac{l_c}{3EI_c} \quad d_{e1} = \frac{l_e}{3EI_e}$$

$$d_{ci} = \frac{l_c}{6EI_c} \quad d_{ei} = \frac{l_e}{6EI_e}$$

Fig. 2.40 Continuous beam
with two spans

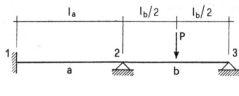

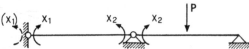

Fig. 2.41 Superimposition
of effects

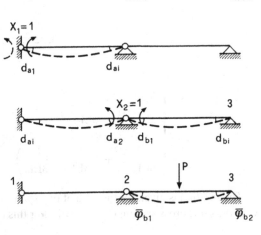

For forces P applied at mid-span, the term due to loads is composed with the
rotations already deduced at the same Sect. 2.1.2:

$$\overline{\varphi}_{c2} = \frac{P_c l_c^2}{16 EI_c} \quad \overline{\varphi}_{e1} = \frac{P_e l_e^2}{16 EI_e}$$

For other types of loads, one can use the pertinent few expressions in good extent
already deduced in Sect. 2.4.

A simple application of the procedure described above is now presented, extended
up to the drawing of the diagrams of internal forces. A continuous beam with two
spans is described in Fig. 2.40, together with the correspondent auxiliary beam chosen
following the quoted standard criterion.

Figure 2.41 shows the superposition of effects for the formulation of the system of compatibility equations that, with the well-known recurrent expressions of
the flexibilities and rotations due to loads, can be directly written summing up the
contributions of relative rotation at the involved joints 1 and 2:

Fig. 2.42 Diagrams of
internal forces

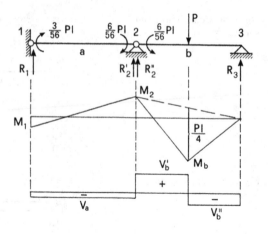

$$\begin{cases} \dfrac{l_a}{3EI_a}X_1 + \dfrac{l_a}{6EI_a}X_2 = 0 \\[2ex] \dfrac{l_a}{6EI_a}X_1 + \left(\dfrac{l_a}{3EI_a} + \dfrac{l_b}{3EI_b}\right)X_2 + \dfrac{Pl_b^2}{16EI_b} = 0 \end{cases}$$

The solution is given for the case of two spans of the same length ($l_a = l_b = l$) and with the same cross section ($I_a = I_b = I$). For this case, the system can be reduced to

$$\begin{cases} 2X_1 + X_2 = 0 \\[1ex] X_1 + 2X_2 = -\dfrac{3}{8}Pl \end{cases}$$

and the solution is

$$X_1 = +\frac{3}{56}Pl$$
$$X_2 = -\frac{6}{56}Pl$$

Introducing these values of the redundant moments in the auxiliary beam, one can calculate the internal forces for each span separately. As shown in Fig. 2.42, it can be set:

$$R_1 = -\frac{X_1 - X_2}{l} = -\frac{9}{56}P \qquad \text{(oriented in downwards direction)}$$

$$R_2 = R_2' + R_2'' = \frac{X_1 - X_2}{l} + \left(-\frac{X_2}{l} + \frac{P}{2}\right)$$
$$= +\frac{9}{56}P + \frac{34}{56}P = +\frac{43}{56}P \qquad \text{(oriented in upwards direction)}$$

$$R_3 = +\frac{X_2}{1} + \frac{P}{2} = +\frac{22}{56}P \quad \text{(oriented in upwards direction)}$$

$$M_1 = X_1 = +\frac{3}{56}Pl \quad \text{(tensioned the lower fibres)}$$

$$M_2 = X_2 = -\frac{6}{56}Pl \quad \text{(tensioned the upper fibres)}$$

$$M_b = +\frac{X_2}{2} + \frac{Pl}{4} = +\frac{11}{56}Pl \quad \text{(tensioned the lower fibres)}$$

$$V_a = R_1 = -\frac{9}{56}P \quad \text{(counter-clockwise)}$$

$$V_b' = R_2'' = +\frac{34}{56}P \quad \text{(clockwise)}$$

$$V_b'' = -R_{23} = -\frac{22}{56}P \quad \text{(counter clockwise)}$$

The compatibility conditions of the Force Method can be set also on an auxiliary structure still hyperstatic. The choice of an isostatic auxiliary structure is suggested by the possibility of an elementary calculation of the coefficients of the equation system. Actually, these coefficients, thanks to the standard procedure, are previously defined and listed in the pertinent tables, without the need of any calculation. If these tables are extended so to contain the flexibilities and rotations due to loads also of the beam with a full support at one end and a simple bearing at the other end, one can reduce the unknowns, with useful simplification of a manual elaboration of the solution of the equations.

Actually, the beam with full support and simple bearing has already been treated at Sect. 2.2.1, obtaining a flexibility

$$d = \frac{l}{4EI}$$

while for other types of loads the end rotations can be deduced from the examples presented at Sect. 2.4.1.

Without the disconnection of the rotation constraint at support 1, the auxiliary beam is arranged as shown in Fig. 2.43, on which one only compatibility equation is imposed:

$$\left(\frac{l_a}{4EI_a} + \frac{l_b}{3EI_b}\right)X_2 + \frac{Pl_b^2}{16EI_b} = 0$$

that, for $l_a = l_b = l$ and $I_a = I_b = I$, gives immediately the solution

Fig. 2.43 Procedure with
hyperstatic auxiliary beam

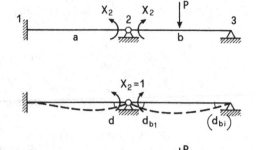

$$X_2 = -\frac{6}{56}Pl = M_2$$

The moment at the end 1, as already found in the quoted example of Sect. 2.2.1 (see
Fig. 2.26), is obtained with

$$M_1 = -X_2/2 = -M_2/2$$

2.2.3   *Elastic Deformable Supports*

The elastic deformable supports are represented, in the compatibility equations of the
Force Method, by their pertinent flexibility contributions to be added to those of the
ordinary beams in the formulation of the coefficients. Let's consider, for example, the
beam of Fig. 2.44 where the full support at the end 1 displays a rotational flexibility
characterized by the elastic constant c_φ; this constant measures the rotation of the
support when subjected to a unity moment and therefore is expressed in 1/(Nm). In
the static scheme of Fig. 2.44, the flexibility of concern is simulated by the rotational
spring.

Following the standard criterion specified in the preceding section, the flexural
continuity restraints on the supports 1 and 2 are disconnected and the same unknowns
X_1 and X_2 of the beam of Fig. 2.40 are applied. The only difference with respect to
the compatibility equations written for that beam lays in the effect of the unknown
X_1 that, further to the rotation d_{a1} shown in Fig. 2.41, produces a rotation c_φ in
the contiguous spring. Adding this contribution and setting $P_a = 0$ and $P_b = P$ for a
direct comparison between the two cases under examination, one obtains the equation
system

$$\begin{cases} \left(c_\varphi + \dfrac{l_a}{3EI_a}\right)X_1 + \dfrac{l_a}{6EI_a}X_2 = 0 \\[3mm] \dfrac{l_a}{6EI_a}X_1 + \left(\dfrac{l_a}{3EI_a} + \dfrac{l_b}{3EI_b}\right)X_2 + \dfrac{Pl_b^2}{16EI_b} = 0 \end{cases}$$

With the chosen sign conventions, that assume positive the rotations in the same directions of the corresponding hyperstatic moments, the additional contribution due to the spring is positive, as always results for all the "direct effects". One can also notice that the support flexibility may be simulated by an "elastic element" interposed between the beam end and the external full support. This element, when subjected to the moment transmitted from the beam, reacts in a way similar to the beams, displaying a rotation proportional to the moment itself. The only difference is the value of the elastic constant (c_φ instead of l/3EI). For the rest, this constant is treated as an ordinary direct flexibility.

The analysis of the beam of Fig. 2.44 is completed with $l_a = l_b = l$ and $I_a = I_b = I$ and assuming for the spring the same flexibility

$$c_\varphi = \frac{l}{3EI}$$

of the contiguous beam, so that the arrangement can correspond to the symmetric beam of Fig. 2.45.

The compatibility equation system becomes

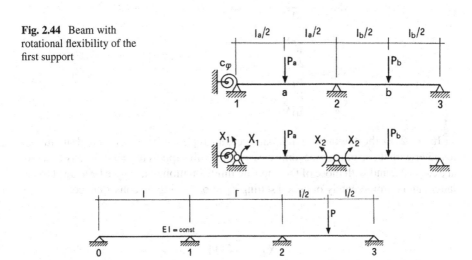

Fig. 2.44 Beam with rotational flexibility of the first support

Fig. 2.45 Equivalent symmetric beam

$$\begin{cases} 4X_1 + X_2 = 0 \\ X_1 + 2X_2 = -\dfrac{3}{8}Pl \end{cases}$$

leading to the solution

$$X_1 = +\frac{1}{40}Pl$$

$$X_2 = -\frac{4}{40}Pl$$

With calculations similar to those performed for the definition of the diagrams of Fig. 2.42, one obtains

$$R_1 = -\frac{5}{40}P$$

$$R_2 = R_2' + R_2'' = +\frac{5}{40}P + \frac{24}{40}P = +\frac{29}{40}P$$

$$R_3 = +\frac{16}{40}P$$

$$M_1 = +\frac{1}{40}Pl$$

$$M_2 = -\frac{4}{40}Pl$$

$$M_b = +\frac{8}{40}Pl$$

$$V_a = -\frac{5}{40}P$$

$$V_b' = +\frac{24}{40}P$$

$$V_b'' = -\frac{16}{40}P$$

In Fig. 2.46, the diagrams of internal forces are reported for the examined situation, together with those of the preceding case of fixed full support ($c_\varphi = 0$) already treated at Sect. 2.2.2 and with those of the opposite limit situation of a simple bearing. These latter can be immediately obtained setting $c_\varphi = \infty$ having by consequence:

$$X_1 = 0$$

$$X_2 = -\frac{3}{32}Pl$$

One can notice the small influence that the constraint degree of the distant support at the end 1 has on the loaded span.

Different is the case of a load applied in the first span of the beam ($P_a = P$ and $P_b = 0$). Figure 2.47 shows this load condition applied to the auxiliary beam, with indicated the end rotations due to the load:

$$\overline{\varphi}_{a1} = \overline{\varphi}_{a2} = \frac{Pl^2}{16EI}$$

Indicating with

$$\kappa_\varphi = \frac{c_\varphi}{1/3EI}$$

the contribution of rotational flexibility of the spring divided by the beam flexibility, the compatibility equation system becomes

Fig. 2.46 Diagrams of internal forces for P on the second span

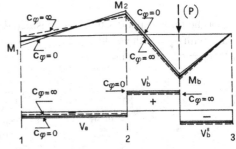

Fig. 2.47 Diagrams of internal forces for P on the first span

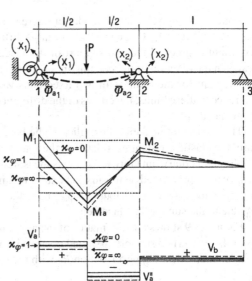

Fig. 2.48 Beam with a
translation flexibility at the
internal support

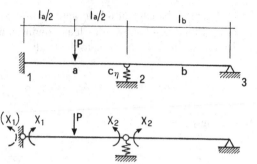

$$\begin{cases} 2(\kappa_\varphi + 1)X_1 + X_2 = -\dfrac{3}{8}Pl \\[2mm] X_1 + 4X_2 = -\dfrac{3}{8}Pl \end{cases}$$

For the three support situations considered above, the one with a fixed end constraint with $\kappa_\varphi = 0$, the middle one with $\kappa_\varphi = 1$ and the one without rotation constraint with $\kappa_\varphi = \infty$, one obtains

κ_φ	0	1	∞
X_1	$-9Pl/56$	$-3Pl/40$	0
X_2	$-3Pl/56$	$-3Pl/40$	$-3Pl/32$

leading, through the same calculations here not reproduced, to the internal forces diagrams of Fig. 2.47. One can notice the large differences consequent to the different constraint degrees of the end support of the loaded span. The dotted straight lines drawn in the diagram of the bending moment indicate the limits $\pm Pl/8$ within which the diagram for the beam with two fixed ends would remain (see Fig. 2.33). With reference to these limits, the effects of the different degrees of flexibility can be better appreciated.

The case of a translation flexibility of the bearings is shown in Fig. 2.48. In this case, the elastic constant c_η, related to the linear spring of the model, expresses the displacement due to an unity force and is measured in m/N. The stress analysis of the beam can be performed on the auxiliary structure, made isostatic in the standard way, adding to the rotation contributions φ'_{jk} due to the span flexibilities the contributions φ^*_{jk} due to the support displacement.

Figure 2.49 shows the kinematics of the auxiliary structure due to a displacement η_2 of the concerned support, assumed positive if directed downwards and expressed as a function of the support reaction R_2 with the elasticity relation

$$\eta_2 = c_\eta R_2$$

Fig. 2.49 Beam kinematics due to translation η_2

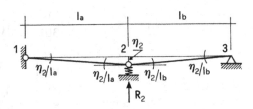

Fig. 2.50 Superposition of reaction effects

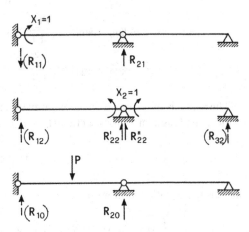

On the joints 1 and 2, this displacement produces the relative rotations:

$$\varphi_1^* = +\frac{\eta_2}{l_a} = +c_\eta \frac{1}{l_a} R_2$$

$$\varphi_2^* = -\frac{\eta_2}{l_a} - \frac{\eta_2}{l_b} = -c_\eta \left(\frac{1}{l_a} + \frac{1}{l_b}\right) R_2$$

In order to introduce these contributions in the compatibility conditions, the support reaction R_2 should be expressed as a function of the hyperstatics and the load. Dealing with an isostatic structure, the support translation η_2 does not affect this reaction. Figure 2.50 shows these contributions separately. Summed up, they give the reaction

$$R_2 = +\frac{1}{l_a} X_1 - \left(\frac{1}{l_a} + \frac{1}{l_b}\right) X_2 + \frac{1}{2} P$$

After this preliminary calculation, the coefficients of the compatibility equations

$$\begin{cases} \varphi_{11} X_1 + \varphi_{12} X_2 + \varphi_{10} = 0 \\ \varphi_{21} X_1 + \varphi_{22} X_2 + \varphi_{20} = 0 \end{cases}$$

can be expressed, through the obvious substitutions, as

$$\varphi_{11} = \varphi'_{11} + \varphi^*_{11} = \frac{l_a}{3EI_a} + c_\eta \left(\frac{1}{l_a}\right)^2$$

$$\varphi_{12} = \varphi'_{12} + \varphi^*_{12} = \frac{l_a}{6EI_a} - \frac{c_\eta}{l_a}\left(\frac{1}{l_a} + \frac{1}{l_b}\right) = \varphi_{21}$$

$$\varphi_{22} = \varphi'_{22} + \varphi^*_{22} = \frac{l_a}{3EI_a} + \frac{l_b}{3EI_b} + c_\eta \left(\frac{1}{l_a} + \frac{1}{l_b}\right)^2$$

$$\varphi_{10} = \varphi'_{10} + \varphi^*_{10} = \frac{Pl_a^2}{16EI_a} + c_\eta \left(\frac{1}{l_a}\right)\frac{P}{2}$$

$$\varphi_{20} = \varphi'_{20} + \varphi^*_{20} = \frac{Pl_a^2}{16EI_a} - c_\eta \left(\frac{1}{l_a} + \frac{1}{l_b}\right)\frac{P}{2}$$

The analysis of the beam under examination is completed in the case of two spans of the same length and cross section ($l_a = l_b = l$ and $I_a = I_b = I$). Assuming

$$\kappa_\eta = \frac{c_\eta}{l^3/3EI}$$

one has the equation system

$$\begin{cases} 2(1 + \kappa_\eta)X_1 + (1 - 4\kappa_\eta)X_2 = -\frac{3}{8}Pl\left(1 + \frac{8}{3}\right)\kappa_\eta \\ \\ (1 - 4\kappa_\eta)X_1 + 4(1 + 2\kappa_\eta)X_2 = -\frac{3}{8}\left(1 - \frac{16}{3}\right)\kappa_\eta \end{cases}$$

that leads to the solution

$$X_1 = -\frac{Pl}{8}3\frac{3 + 28\kappa_\eta}{7 + 32\kappa_\eta} = Xc_1$$

$$X_2 = -\frac{Pl}{8}\frac{3 - 22\kappa_\eta}{7 + 32\kappa_\eta} = Xc_2$$

where $X = -Pl/8$ is the moment reaction that would arise at the end supports of the loaded span if they were fixed full restraints.

Figure 2.51 gives the variation curves of the two hyperstatic moments as a function of the flexibility ratio κ_η. They start, for a null flexibility, from the values

$$X_1 = -\frac{9}{56}Pl = 1.286X$$

$$X_2 = -\frac{3}{56}Pl = 0.429X$$

already obtained for the preceding case (see curve $\kappa_\varphi = 0$ of Fig. 2.47). As the flexibility increases, the moment at the end 1 increases, while the moment at the

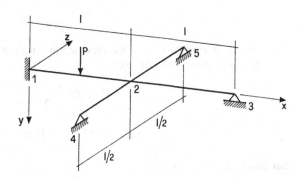

Fig. 2.51 Variation curves of the hyperstatic moments

Fig. 2.52 Structural situation with supporting and supported beams

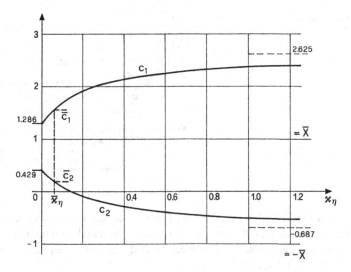

continuity support 2 decreases so much as to change its sign. For $\kappa_\eta \to \infty$ the values tend to those of a beam with only one span long $2l$ without the middle support:

$$X_1 = -\frac{Pl}{8} 3 \frac{28}{32} = 2.625X$$

$$X_2 = -\frac{Pl}{8} \frac{-22}{32} = -0.687X$$

In order to show the magnitude order of the concerned effects in a real structural situation such as that represented in Fig. 2.52, let's consider the beam 1–2–3 under examination laid at its ends on columns that, due to their axial high stiffness, ensure a null vertical translation of the supports. The intermediate bearing instead is given by a transverse supporting beam 4–2–5 with a flexural flexibility similar to that of the supported beam 1–2–3.

Fig. 2.53 Diagrams of
internal forces

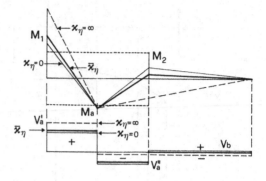

The measures indicated in Fig. 2.52 are taken, assuming all the longitudinal and
transverse beams made with the same cross section. The translation flexibility of the
middle joint of the transverse beam 4–5 can be deduced from the case of Fig. 2.15
presented at Sect. 2.1.2 setting $P = 1$:

$$c_\eta = \delta_0 = \frac{l^3}{48EI}$$

This is the flexibility of the intermediate bearing of the supported beam 1–2–3.
With these values, one obtains:

$$\bar{\kappa}_\eta = \frac{l^3/48EI}{l^3/3EI} = 0.0625$$

$$\bar{c}_1 = 1.583 \quad \bar{c}_2 = 0.181$$

that lead to the internal forces:

$$M_1 = X_1 = -1.583\frac{Pl}{8}$$

$$M_2 = X_2 = -0.181\frac{Pl}{8}$$

$$M_a = \frac{X_1 + X_2}{2} + \frac{Pl}{4} = +1.118\frac{Pl}{8}$$

$$V'_a = +R_1 = -\frac{X_1 - X_2}{1} + \frac{P}{2} = +0.675P$$

$$V''_a = +R'_2 = -\frac{X_1 - X_2}{1} - \frac{P}{2} = -0.325P$$

$$V_b = -R_3 = -\frac{X_2}{1} = +0.023P$$

Figure 2.53 shows the related diagrams together with those of the two limit situations examined before. One can notice the high influence of the support flexibility

on the internal forces and how also ordinary structural situations can lead to relevant differences with respect to what deducible from an approximate scheme that neglects the support flexibility.

2.3 Application to Frames

The Force Method, as presented in the preceding section, can shorten and facilitate the analysis of hyperstatic structures with respect to other methods, such as the integration of the elastic line or the principle of virtual works. When applied to frames, it finds limits of efficiency for which in most cases it is less convenient that the method presented in the following chapter.

Actually, its convenience depends on the possibility of a standardized choice of the auxiliary structure so that the definitions of the coefficients of the solving equations become ready. This possibility holds good only if, applying integrally the procedure of disconnection of the flexural continuity restraints at the joints of the frame, it does not become hypostatic and the effects of the axial deformations of the members on the internal forces remain null or negligible. On the contrary case, the calculation of the coefficients of the compatibility equations requires, instead of the separate analysis of any single beam, the global analysis of the structure repeated on the basis of the kinematics proper of the specific structural arrangement.

What stated above will be better clarified through the examples presented in the following sections on which the possibility of some important simplifications of the analysis will be discussed. Only some few definitions are now set, made necessary by the transfer, from the one-dimensional problem of the beam in flexure to the frames arranged in the two-dimensional plane or even in the three-dimensional space.

Similar to the flexural deformation parameters presented at Sect. 2.1 (see Fig. 2.11 with the direct d_1, d_2 and indirect d_i flexibilities), the *axial deformability* of a straight beam is defined as the elongation produced in it by an unity force (see Fig. 2.54a). Integrating the elementary deformations

$$du = \frac{N}{EA} d\xi = \varepsilon_x d\xi,$$

with $N = 1$, one has

$$d_x = \int_0^l \frac{1}{EA(\xi)} d\xi \simeq \sum_{i=1}^n \frac{\Delta\xi}{EA_i}$$

that, for a constant cross section ($A = \text{const.}$), becomes

$$d_x = \frac{l}{EA}$$

Fig. 2.54 Axial (**a**) and torsional (**b**) behaviour of a beam

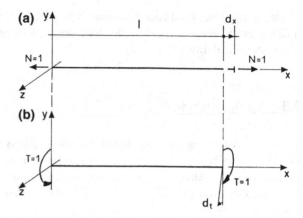

The torsional behaviour of the beam (Fig. 2.54b) is deduced in a similar way defining a *torsional deformability* d_t as the relative rotation between the two end sections produced by an unity torsional moment. Integrating the elementary deformations of the uniform torsion:

$$d\varphi_x = \frac{T}{GJ}d\xi = \chi_x d\xi,$$

with $T = 1$, one has

$$d_t = \int_0^1 \frac{1}{GJ(\xi)}d\xi \simeq \sum_{i=1}^n \frac{\Delta\xi}{GJ_i}$$

that, for a constant cross section ($J = $ const.), becomes

$$d_t = \frac{1}{GJ}$$

The symbol J indicates the "torsional second moment of area" of the section; its value depends on the shape of the section itself.

2.3.1 Contribution of Axial Deformation

Let's consider the frame of Fig. 2.55a, and for its analysis, let's apply the criteria of the standard procedure of the Force Method disconnecting the flexural continuity restraints at the joints 1 and 3. On the isostatic auxiliary structure obtained in this way (see Fig. 2.55b), the hyperstatic moments X_1 and X_3 are applied in addition to the given load P.

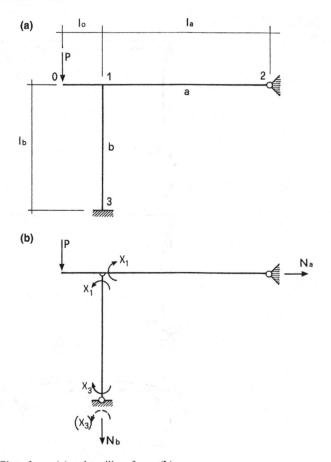

Fig. 2.55 Given frame (**a**) and auxiliary frame (**b**)

For the compatibility equation system

$$\begin{cases} \varphi_{11}X_1 + \varphi_{13}X_3 + \varphi_{10} = 0 \\ \varphi_{31}X_1 + \varphi_{33}X_3 + \varphi_{30} = 0 \end{cases}$$

the coefficients are composed of a flexural part φ'_{jk} and of a kinematic part φ^*_{jk}: The first one is deduced from flexural behaviour of the members as shown in Fig. 2.56; the second comes from the axial deformation of the members as shown in Fig. 2.57.

The flexural contribution, referring again to the beams hinged at the ends and considering separately the translations of joint 1, is evaluated with the same expressions of the continuous beams:

$$\varphi'_{11} = \frac{l_a}{3EI_a} + \frac{l_b}{3EI_b}$$

Fig. 2.56 Deduction of the
flexural part of the
coefficients

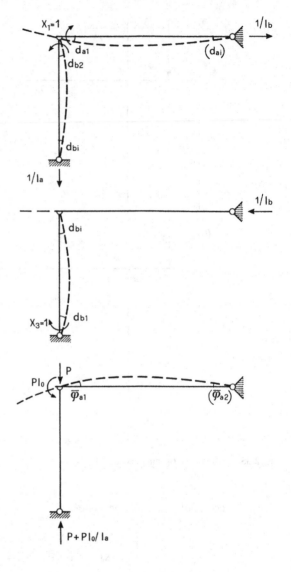

$$\varphi'_{13} = \frac{l_b}{6EI_b} = \varphi'_{31}$$

$$\varphi'_{33} = \frac{l_b}{3EI_b}$$

$$\varphi'_{10} = \frac{-l_a}{3EI_a}Pl_o$$

$$\varphi'_{30} = 0$$

Fig. 2.57 Contributions of
the axial deformations

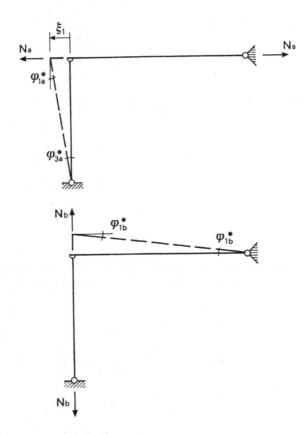

In particular, it can be noticed how the action of the given load has been brought
to the joint 1, taking away the isostatic cantilever and entering in this way the set
of elementary cases of a simply supported beam already available in the pertinent
tables.

For each of the three situations of Fig. 2.56, the one with the first hyperstatic made
unitary, the one with the second hyperstatic made unitary and the one with the given
load, the axial actions on the two involved members are also indicated. Summing up
these actions, that are assumed positive if in tension, one has:

$$N_a = +\frac{1}{l_b}X_1 - \frac{1}{l_b}X_3 + 0$$
$$N_b = +\frac{1}{l_a}X_1 + 0 - \frac{l_a+l_o}{l_a}P$$

The translations of the joint 1 due to the elongation of the members are indicated
in Fig. 2.57 and are:

$$\xi_1 = \frac{l_a}{EA_a} N_a$$

$$\eta_1 = \frac{l_b}{EA_b} N_b$$

These translations lead, respectively, to the rotations:

$$\varphi_{1a}^* = +\frac{\xi_1}{l_b} = +\frac{l_a}{l_b}\frac{N_a}{EA_a}$$

$$\varphi_{3a}^* = -\frac{\xi_1}{l_b} = -\frac{l_a}{l_b}\frac{N_a}{EA_a}$$

$$\varphi_{1b}^* = +\frac{\eta_1}{l_a} = +\frac{l_b}{l_a}\frac{N_b}{EA_b}$$

$$\varphi_{3b}^* = 0$$

that, as already said, are to be added to the flexural ones in the formulation of the equation system. That is one has, after the pertinent substitutions:

$$\varphi_{11} = \varphi_{11}' + \varphi_{11}^* = \frac{l_a}{3EI_a} + \frac{l_b}{3EI_b} + \frac{1}{l_b^2}\frac{l_a}{EA_a} + \frac{1}{l_a^2}\frac{l_b}{EA_b}$$

$$\varphi_{13} = \varphi_{13}' + \varphi_{13}^* = \frac{l_b}{6EI_b} + \frac{1}{l_b^2}\frac{l_a}{EA_a} = \varphi_{31}$$

$$\varphi_{33} = \varphi_{33}' + \varphi_{33}^* = \frac{l_b}{3EI_b} + \frac{1}{l_b^2}\frac{l_a}{EA_a}$$

$$\varphi_{10} = \varphi_{10}' + \varphi_{10}^* = -\frac{l_a}{3EI_a}Pl_o - \frac{l_o + l_a}{l_a}\frac{l_b}{EA_b}P$$

$$\varphi_{30} = \varphi_{30}' + \varphi_{30}^* = 0$$

To show the magnitude order of the influence of the axial deformation of the members on the internal forces, let's set $l_a = l_b = 1$, $l_a = 4l_o$, $A_a = A_b = A$ and $I_a = I_b = I$. Defining

$$\kappa_o = \frac{I}{l^2 A}$$

the coefficient related to the slenderness of the members, the compatibility equation system becomes

$$\begin{cases} 4(1 + 3\kappa_o)X_1 + (1 - 6\kappa_o)X_3 = 2Pl_o(1 + 15\kappa_o) \\ (1 - 6\kappa_o)X_1 + 2(1 + 3\kappa_o)X_3 = 0 \end{cases}$$

It leads to the solution:

$$X_1 = +\frac{4Pl_o\kappa_p\kappa}{8\kappa^2 - \kappa_i} = +\frac{4}{7}Pl_o\kappa_1$$

$$X_3 = -\frac{2Pl_o\kappa_p\kappa_i}{8\kappa^2 - \kappa_i} = -\frac{2}{7}Pl_o\kappa_3$$

having set

$$\kappa = 1 + 3\kappa_o$$

$$\kappa_i = 1 - 6\kappa_o$$

$$\kappa_p = 1 + 15\kappa_o$$

At the limit of the axial undeformability, with $A \to \infty$ and $\kappa_o \to 0$, it remains

$$\kappa_1 = \frac{7\kappa_p\kappa}{8\kappa^2 - \kappa_i^2} = 1$$

$$\kappa_3 = \frac{7\kappa_p\kappa_i}{8\kappa^2 - \kappa_i^2} = 1$$

values to which tend those of the real structural situations as the member slenderness increases. Let's consider, for example, the two types of sections, rectangular and I shaped, of Fig. 2.19. For these sections, indicating with h their depth, one obtains, respectively:

$$\kappa_o = \frac{Ah^2/12}{Al^2} = \frac{1}{12\lambda^2}$$

$$\kappa_o \simeq \frac{0.67Ah^2/4}{Al^2} = \frac{1}{6\lambda^2}$$

The curves of Figs. 2.58 and 2.59 are consequently deduced in which the field of the deep beams is evidenced, for which the effects of the axial deformations are sensible. For the higher slenderness instead, these effects remain small, and this allows to neglect, in the analysis of the frame, the kinematic contributions of Fig. 2.57.

What stated above shall be considered as a general indication due to the approximations made. Considering also the similar conclusions about the influence of the shear deformation (see Sect. 2.1.3), one can indicate with

$$\lambda = \frac{1}{h} > 5$$

the field of the slenderness in which the approximate procedures of frame analysis presented in these chapters can be applied. Besides the small influence of the shear and axial deformations, in this field of slender beams, other approximations of the calculation model remain admissible such as those related to the neglected actual finite dimensions of the structural joints (see Fig. 1.14 of Sect. 1.1.2).

Fig. 2.58 Curves of the influence of axial deformations—rectangular section

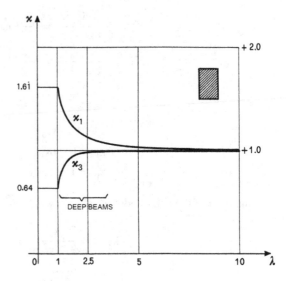

Fig. 2.59 Curves of the influence of axial deformations—I-shaped section

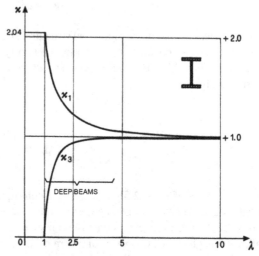

There are anyway some types of structures, like the supporting tripods of big machinery or other structures for industrial plants, which shape ratios do not allow these approximations.

For completeness' sake in Fig. 2.60, the diagrams of internal forces are reported with reference to the approximate solution that neglects the kinematic contributions and for which one has:

$$M_o = Pl_o$$

$$M_{a1} = -Pl_o + X_1 = -\frac{3}{7}Pl_o$$

Fig. 2.60 Diagrams of internal forces

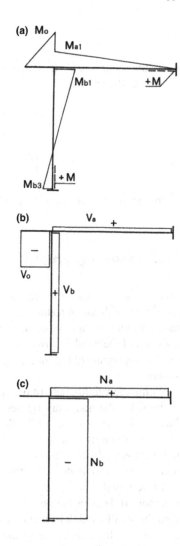

where the bending moments are assumed positive if they tension the fibres outlined in Fig. 2.60a, and again:

$$V_o = -P$$

$$V_a = -\frac{M_{a1}}{1} = +\frac{3}{28}P$$

$$V_b = -\frac{M_{b3} - M_{b1}}{1} = +\frac{6}{28}P$$

where the shear force is assumed positive if clockwise ($1 = 4l_o$).

Eventually, one has:

$$N_a = \frac{X_1}{1} - \frac{X_3}{1} = +\frac{6}{28}P \quad (=V_b)$$

$$N_b = \frac{X_1}{1} - \frac{1+l_o}{1}P = -\frac{31}{28}P \ (=V_o - V_a)$$

where the axial force is assumed positive if of tension.

2.3.2 Non-sway Frames

There are frames for which the auxiliary structure deduced, following the standard procedure, with the introduction of hinges in all the joints does not become hypostatic. If there are not elastic deformable translation supports, for these frames, the omission of the axial deformations of the members leads to consider null the translation components of the joint displacements. This category is defined as *non-sway frames*.

Within the validity field of the approximate analysis that neglects the axial deformations, for the non-sway frames, it is possible to apply the Force Method using the same few expressions of the flexibilities of the doubly supported beam as made for the continuous beams. In this way, the flexural behaviour of bending moment and shear is analysed. The axial forces can be subsequently deduced on the isostatic auxiliary structure under the action of loads and hyperstatics.

To be noted that for the frames the axial forces do not influence the flexural behaviour if, besides the possibility to neglect the kinematic contributions deriving from the small axial deformations, one remains in the field of the first-order theory that neglects the deflections of the members with respect to the action line of the applied forces. So at one side, the deep beams are excluded, and at the other side also, the high slenderness are excluded. For these high slendernesses of beams, the procedures of structural analysis should be adapted following the criteria of the second-order theory that will be thoroughly treated in Chap. 4.

Let's consider now the frame of Fig. 2.61a, for which Fig. 2.61b shows the auxiliary structure chosen on the basis of the standard procedure. This latter does not become hypostatic, and so the given one is a non-sway frame.

For what concerns the hyperstatic moments to be evidenced, consistently with the flexural continuities that have been disconnected, in each joint, these are as many as the converging members less one. So in the joint 1 of the frame of Fig. 2.61, the hyperstatic moment X_1 is sufficient to represent the continuity between the members c and a; on joint 2 instead two hyperstatic moments should be evidenced, as shown in

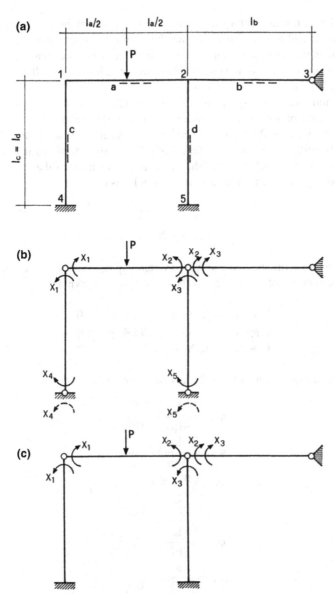

Fig. 2.61 Non-sway frame—solution procedure

the figure, where the moment X_2 has been chosen to represent the continuity between the members a and b, the moment X_3 has been chosen to represent the continuity between the members d and b. One has to remember that in these cases, in order to avoid confusion of symbols, the numbering of hyperstatics cannot follow closely the numbering of joints.

For the non-sway frames, it is possible also to take profit of the equation saving deriving from the omission of the disconnection of the flexural restraints at the external full supports. In Fig. 2.61c, this choice is shown. As already said in Sect. 2.2.2 (see Fig. 2.43), the different expression of the flexibility ($l/4EI$ instead of $l/3EI$) implicitly takes into account the behaviour of the different arrangement of the members with one full end support that for the case of concern leads to:

$$X_4 = -X_1/2$$
$$X_5 = -X_3/2$$

So the three equations of the solving system express the rotation compatibility between the sections of application of the corresponding unknowns:

$$\begin{cases} \varphi_{11}X_1 + \varphi_{12}X_2 + \varphi_{13}X_3 + \varphi_{10} = 0 \\ \varphi_{21}X_1 + \varphi_{22}X_2 + \varphi_{23}X_3 + \varphi_{20} = 0 \\ \varphi_{31}X_1 + \varphi_{32}X_2 + \varphi_{33}X_3 + \varphi_{30} = 0 \end{cases}$$

With the usual expressions, the coefficients are written as:

$$\varphi_{11} = \frac{l_c}{4EI_c} + \frac{l_a}{3EI_a}$$

$$\varphi_{12} = \frac{l_a}{6EI_a} = \varphi_{21}$$

$$\varphi_{13} = 0 = \varphi_{31}$$

$$\varphi_{22} = \frac{l_a}{3EI_a} + \frac{l_b}{3EI_b}$$

$$\varphi_{23} = \frac{l_b}{3EI_b} = \varphi_{32}$$

$$\varphi_{33} = \frac{l_b}{3EI_b} + \frac{l_d}{4EI_d}$$

$$\varphi_{10} = \frac{Pl_a^2}{16EI_a}$$

$$\varphi_{20} = \frac{Pl_a^2}{16EI_a}$$

$$\varphi_{30} = 0$$

The solution is elaborated for $l_a = l_b = l_c = l_d = 1$, $I_a = I_b = 2I$ and $I_c = I_d = I$. The corresponding equation system becomes

$$\begin{cases} 5X_1 + X_2 = -\dfrac{3}{8}Pl \\[2mm] X_1 + 4X_2 + X_3 = -\dfrac{3}{8}Pl \\[2mm] X_2 + 5X_3 = 0 \end{cases}$$

and leads to

$$X_1 = -\frac{7}{120}Pl$$
$$X_2 = -\frac{10}{120}Pl$$
$$X_3 = +\frac{2}{120}Pl$$

With this solution, one has the isostatic structure of Fig. 2.62, on which with simple calculations one has the values:

$$M_{a1} = X_1 = -\frac{14}{240}Pl$$
$$M_{a2} = X_2 = -\frac{20}{240}Pl$$
$$M_a = \frac{X_1 + X_2}{2} + \frac{Pl}{4} = +\frac{43}{240}Pl$$
$$M_{b2} = X_2 + X_3 = -\frac{16}{240}Pl$$
$$M_{c4} = -\frac{X_1}{2} = +\frac{7}{240}Pl$$

Fig. 2.62 Isostatic structure after solution

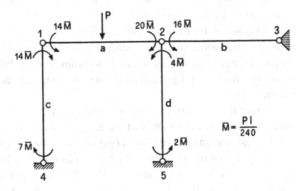

$$M_{c1} = X_1 = -\frac{14}{240}Pl$$

$$M_{d5} = -\frac{X_3}{2} = -\frac{2}{240}Pl$$

$$M_{d2} = X_3 = +\frac{4}{240}Pl$$

and subsequently:

$$V_a' = +\frac{P}{2} - \frac{M_{a1} - M_{a2}}{1} = +\frac{114}{240}P$$

$$V_a'' = -\frac{P}{2} - \frac{M_{a1} - M_{a2}}{1} = -\frac{126}{240}P$$

$$V_b = -\frac{M_{b2}}{1} = +\frac{16}{240}P$$

$$V_c = -\frac{M_{c4} - M_{c1}}{1} = -\frac{21}{240}P$$

$$V_d = -\frac{M_{d5} - M_{d2}}{1} = +\frac{6}{240}P$$

$$N_a = V_c = -\frac{21}{240}P$$

$$N_b = N_a + V_d = -\frac{15}{240}P$$

$$N_c = -V_a' = -\frac{114}{240}P$$

$$N_d = +V_a'' - V_b = -\frac{142}{240}P$$

From all these values, expressed with the same sign conventions specified before, the diagrams of Fig. 2.63 are obtained.

Independently from the flexural analysis that can be performed, with the criteria presented above, examining separately each member of the auxiliary structure, the axial forces system may present some redundancy to be solved with the addition of pertinent compatibility conditions.

Let's assume, for example, that the frame just analysed was provided also with a translation restraint at joint 1 as described in Fig. 2.64a. Nothing changes in this case for the flexural analysis that would lead again to the solution of Fig. 2.64b. But this auxiliary structure still presents a redundancy with respect to the axial behaviour of beams 1–2–3.

In order to solve this redundancy, one may proceed following the criteria of the Force Method, assuming an isostatic auxiliary beam with the disconnection at joint 3 of the translation restraint, and applying on it the corresponding hyperstatic force X. This is shown in Fig. 2.64c where the columns have been taken off and replaced by their known actions V_c and V_d. Also, the transverse actions have been omitted since they do not affect the axial behaviour under examination.

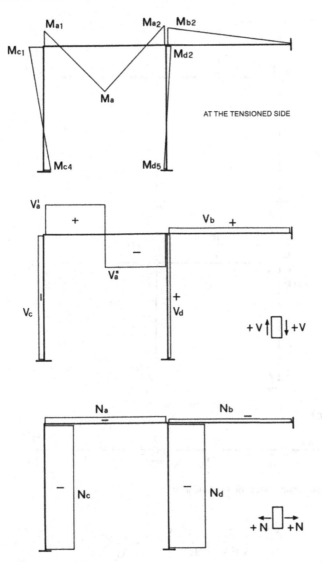

Fig. 2.63 Diagrams of internal forces

The compatibility condition imposes a zero value to the translation of joint 3:

$$\xi = 0$$

and is written on the basis of the axial deformability of the two involved members:

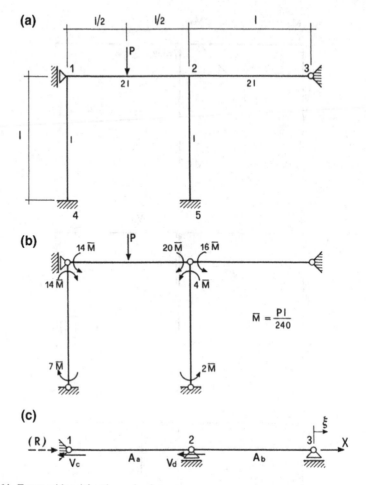

Fig. 2.64 Frame with axial action redundancy

$$\left(\frac{l_a}{EA_a} + \frac{l_b}{EA_b}\right)X - \frac{l_a}{EA_a}V_d = 0$$

This leads to the solution

$$X = \frac{l_a/A_a}{l_a/A_a + l_b/A_b}V_d$$

With $l_a = l_b = l$ and $A_a = A_b = A$, for $V_d = 6P/240$ and $V_c = -21P/240$, one obtains, for example:

$$X = \frac{1}{2}\frac{6P}{240}$$

$$N_b = X = +\frac{3}{240}P$$

$$N_a = X - V_d = -\frac{3}{240}P$$

$$R = -N_a + V_c = -\frac{18}{240}P$$

and this defines the repartition of the horizontal forces acting on the beam between the two end supports 1 and 3.

2.3.3 Sway Frames

Are defined as *sway frames* the ones for which the auxiliary structure, deduced with the introduction of hinges in all the joints, becomes hypostatic, or for which there are translation elastic deformability of the supports so that, also if the axial deformations of the members are neglected, the translation components of the joint displacements are not null. For these frames by consequence, the kinematic contributions of the member rotations should be introduced in the compatibility conditions of the Force Method in addition to flexural ones: the former due to the joint translations and the latter due to the member deflections.

The first way to apply the Force Method to the analysis of the sway frames does not follow thoroughly the standard procedure, leaving in the auxiliary structure the rotation constraints necessary to ensure an isostatic arrangement. With this choice, one renounces the possibility to examine separately any single member by means of few recurrent expressions; in fact, any single action causes joint translations and requires the analysis of the consequent overall kinematic displacement.

For an orderly formulation of the compatibility equations, following the quoted application procedure, one can compute separately the flexural "non-sway" contributions from those produced by the joint translations, similar to what performed at Sect. 2.2.3 to express the effects of the elastic linear deformations of the supports of the continuous beams (see Fig. 2.49).

Let's consider the example of a simple portal described in Fig. 2.65a. For this portal, Fig. 2.65b shows the correspondent kinematic hypostatic displacement that would occur if one introduces hinges in all the four joints. Figure 2.66a shows the isostatic auxiliary structure chosen maintaining the full efficiency of the external support of the right column and evidencing the three hyperstatic moments X_1, X_2 and X_3. Figure 2.66b shows the displacement of the structure that arises when, neglected the axial deformations of the members, a translation ξ is applied to the top beam with lateral deflection of the right column.

So interpreting the right column as an "elastic element" that provides the horizontal support to the remaining structure, exchanging with it the force r evidenced in Fig. 2.66b, one can set the compatibility equation system as:

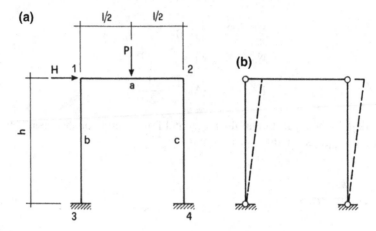

Fig. 2.65 Simple portal sway frame

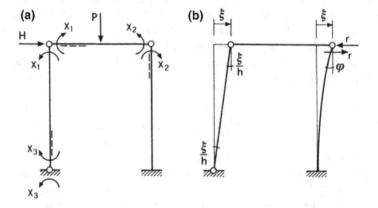

Fig. 2.66 Isostatic auxiliary structure

$$\begin{cases} \left(\dfrac{1}{3EI_a} + \dfrac{h}{3EI_b}\right)X_1 + \dfrac{1}{6EI_a}X_2 + \dfrac{h}{6EI_b}X_3 + \dfrac{Pl^2}{16EI_a} + \varphi_1^* = 0 \\[3mm] \dfrac{1}{6EI_a}X_1 + \dfrac{1}{3EI_a}X_2 + \dfrac{Pl^2}{16EI_a} + \varphi_2^* = 0 \\[3mm] \dfrac{h}{6EI_b}X_1 + \dfrac{h}{3EI_b}X_3 + \varphi_3^* = 0 \end{cases}$$

where the flexural contributions of the part 3–1–2 of the frame have been expressed, without displacements of the elastic support "E.E." at joint 2, while the effects of these displacements have to be added in the terms φ_1^*, φ_2^* and φ_3^*. These terms are evaluated here under (see Fig. 2.67a).

Fig. 2.67 Auxiliary structure with elastic support

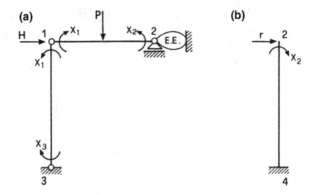

So, with reference to Fig. 2.67b, the rotation φ and the translation ξ of the end 2 of the column are expressed as obtained at Sects. 2.1.1 and 2.1.3 for the cases of Figs. 2.9 and 2.17:

$$\varphi = \frac{h}{EI_c}X_2 + \frac{h^2}{2EI_c}r$$

$$\xi = \frac{h^2}{2EI_c}X_2 + \frac{h^3}{3EI_c}r$$

The value of the action r coming from the other part of the structure is:

$$r = \frac{X_3 - X_1}{h} + H$$

with which, through obvious passages, one obtain

$$\varphi_1^* = -\frac{\xi}{h} = +\frac{h}{3EI_c}X_1 - \frac{h}{2EI_c}X_2 - \frac{h}{3EI_c}X_3 - \frac{Hh^2}{3EI_c}$$

$$\varphi_2^* = +\varphi = -\frac{h}{2EI_c}X_1 + \frac{h}{EI_c}X_2 + \frac{h}{2EI_c}X_3 + \frac{Hh^2}{2EI_c}$$

$$\varphi_3^* = +\frac{\xi}{h} = -\frac{h}{3EI_c}X_1 + \frac{h}{2EI_c}X_2 + \frac{h}{3EI_c}X_3 + \frac{Hh^2}{3EI_c}$$

An alternative way to apply the Force Method to the analysis of the sway frames consists of applying thoroughly the standard procedure and to fix the auxiliary structure with additional translation supports. For the same portal of Fig. 2.65a, one obtain in this way the auxiliary structure of Fig. 2.68a, with five modifications with respect to the given one: with four flexural disconnections at joints 1, 2, 3 and 4 and with the translation support added to the beam 1–2.

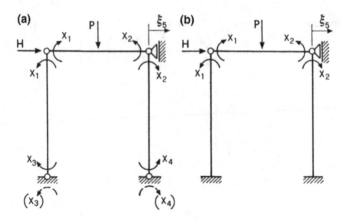

Fig. 2.68 a Standard and **b** improved auxiliary structures

The equivalence between the two structures can be set evidencing first the four hyperstatic moments X_1, X_2, X_3 and X_4 that represent the flexural continuities disconnected; besides these, the translation ξ_5 should be also evidenced to represent the translation freedom of the beam prevented by the additional support. To the four static unknowns correspond as many compatibility conditions of the relative rotations at the joints of the frame; to the geometric unknown instead corresponds the equilibrium condition of the reaction of the additional support. This latter condition expresses the null value of the quoted reaction.

In this way, the solving equation system is:

$$
\begin{cases}
\varphi_1 = 0 \\
\varphi_2 = 0 \\
\varphi_3 = 0 \\
\varphi_4 = 0 \\
r = 0
\end{cases}
$$

where one can notice, together with the non-homogeneity of the equations, partly of compatibility and partly of equilibrium, the high number of equations with respect to the redundancy degree of the problem. In fact, if m is the redundancy degree of the given structure an n is the number of labilities of the auxiliary structure with hinges in all the joints, in addition to the m static unknowns strictly necessary, other n static unknowns are needed corresponding to the disconnections in excess and other n geometric unknowns are needed to restrain the consequent labilities. So the cost payed to apply thoroughly the standard procedure would be high.

The number of equations can be reduced with a different choice of the auxiliary structure as shown in Sect. 2.3.2 for the non-sway frames (see Fig. 2.61c). If, for example, the flexural constraints of the external base supports of the concerned portal are maintained, one can operate on the auxiliary structure of Fig. 2.68b with the only

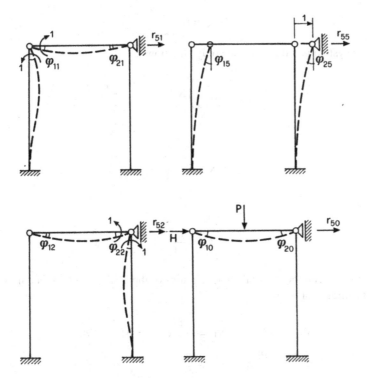

Fig. 2.69 Superposition of effects on the improved auxiliary structure

three unknowns φ_1, φ_2 and ξ_5. Figure 2.69 illustrates the superposition of the effects of the three unknowns and the applied loads from which one has the system:

$$\begin{cases} \left(\dfrac{h}{4EI_b} + \dfrac{1}{3EI_a}\right)X_1 + \dfrac{1}{6EI_a}X_2 - \dfrac{3}{2h}\xi_5 + \dfrac{Pl^2}{16EI_a} = 0 \\[2mm] +\dfrac{1}{6EI_a}X_1 + \left(\dfrac{h}{4EI_c} + \dfrac{1}{3EI_a}\right)X_2 + \dfrac{3}{2h}\xi_5 + \dfrac{Pl^2}{16EI_a} = 0 \\[2mm] +\dfrac{3}{2h}X_1 - \dfrac{3}{2h}X_2 + \left(\dfrac{3EI_b}{h^3} + \dfrac{3EI_c}{h^3}\right)\xi_5 - H = 0 \end{cases}$$

made of two compatibility equations and one equilibrium equation, where the coefficients of the last equation have been deduced from the cases of Figs. 2.29 and 2.24 of Sect. 2.2.1.

The solution of the portal under examination is completed assuming $I_c = I_b = I$ and considering first the *symmetric load condition* with $P > 0$ and $H = 0$ for which it can be set:

$$X_1 = X_2$$

$$\xi = 0$$

so that the system is reduced to one only equation

$$\left(\frac{h}{4EI} + \frac{1}{2EI_a}\right)X_1 = -\frac{Pl^2}{16EI_a}$$

that leads to

$$X_1 = X_2 = -\frac{Pl}{8}\frac{2}{2+\kappa}$$

with

$$\kappa = \frac{h/I}{1/I_a}$$

is the ratio between the flexural characteristics of the column and the beam. So one has the bending moments:

$$M_{a1} = M_{a2} = +X_1 = -\frac{Pl}{8}c_1 \quad (= M_{b1} = M_{c2})$$

$$M_a = +\frac{Pl}{4} + X_1 = +\frac{Pl}{8}c_a$$

$$M_{b3} = M_{c4} = -\frac{X_1}{2} = +\frac{Pl}{16}c_1$$

with

$$c_1 = 2/(2+k) \quad c_a = 2 - c_1$$

The diagrams of internal forces are shown in Fig. 2.70 for three situations: first the limit one for very flexible column ($\kappa \to \infty$), second the limit one with very stiff column ($\kappa \to 0$) and third the intermediate one with the flexibility of the column double than of the beam ($\kappa = 2$). For the bending moments, as a ratio to the "basic" one $M = Pl/8$, one has the values given in the following table.

κ	c_1	c_a
∞	0.0	2.0
2	0.5	1.5
0	1.0	1.0

One can see that the bending moment diagram of the beam remains included between the limit one of two simple supports ($\kappa = \infty$) and the limit one of two full end supports ($\kappa = 0$).

Fig. 2.70 Diagrams of internal forces

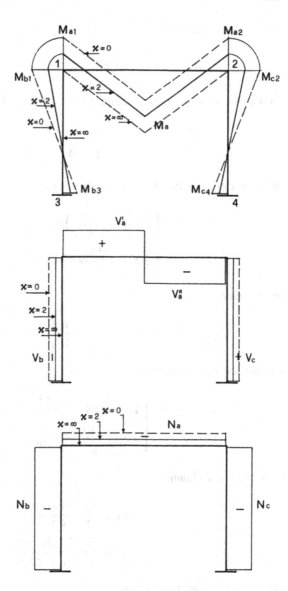

The diagrams of shear and axial force are calculated on the basis of the following equilibria, where it is set $l/h = 4/3$:

$$V'_a = V''_a = +\frac{P}{2}$$

$$V_b = -V_c = +\frac{M_{b1} - M_{b3}}{h} = -\frac{3}{16}P\frac{l}{h}c_1 = -\frac{P}{4}c_1$$

$$N_b = N_c = -\frac{P}{2}$$

$$N_a = +V_b = -\frac{P}{4}c_1$$

For the *anti-symmetric load condition* $P = 0$ and $H > 0$, one has

$$X_1 = -X_2$$
$$\xi \neq 0$$

and the system is reduced to the two equations:

$$\begin{cases} \left(\dfrac{h}{4EI_b} + \dfrac{1}{6EI_a}\right)X_1 - \dfrac{3}{2h}\xi_5 = 0 \\ \dfrac{3}{h}X_1 + \dfrac{6EI_b}{h^3}\xi_5 = H \end{cases}$$

Setting again

$$\kappa = \frac{h/I}{1/I_a}$$

the equation system becomes

$$\begin{cases} (2 + 3\kappa)X_1 - 3\kappa\dfrac{6EI}{h^2}\xi_5 = 0 \\ 3X_1 + \dfrac{6EI}{h^2}\xi_5 = Hh \end{cases}$$

from which one obtains

$$X_1 = \frac{Hh}{2}\frac{6\kappa}{2 + 12\kappa}$$

$$\xi_5 = \frac{Hh}{2}\frac{2 + 3\kappa}{2 + 12\kappa}\frac{h^2}{3EI}$$

arriving to the bending moments

$$M_{a1} = -M_{a2} = +X_1 = +\frac{Hh}{2}c_2 \quad (= M_{b1} = -M_{c2})$$

$$M_{b3} = -M_{c4} = -\frac{X_1}{2} - \frac{3EI}{h^2}\xi_5 = -\frac{Hh}{2}c_3$$

with

$$c_2 = \frac{6\kappa}{2 + 12\kappa}$$

$$c_3 = \frac{2 + 6\kappa}{2 + 12k}$$

Figure 2.71 shows the diagrams of the internal forces for the two limit situations of very stiff beam ($\kappa \to \infty$) and very flexible beam ($\kappa \to 0$) and for the same intermediate situation already considered for the symmetric load condition ($\kappa = 2$). For the bending moments, as a ratio to the "basic" one $M = Hh/2$, one has the values obtainable from the following table.

κ	c_2	c_3
∞	0.5	0.5
2	6/13	7/13
0	0.0	1.0

One can see that the bending moment diagrams of the columns remain included between the limit of the cantilever arrangement ($\kappa = 0$) and the limit of anti-symmetric of two end rotation restraints ($\kappa = \infty$). In Fig. 2.72a, b, these limit situations are represented, with reference to a single column subjected to its part H/2 of the horizontal action, by the corresponding flexural deformations.

The diagrams of shear and axial force are calculated on the basis of the following equilibria, where again it is set $l/h = 4/3$:

$$V_a = -2\frac{X_1}{l} = -\frac{h}{l}Hc_2 = -\frac{3H}{4}c_2$$

$$V_b = V_c = +\frac{H}{2}$$

$$N_a = -\frac{H}{2}$$

$$N_b = -N_c = -V_a = +\frac{h}{l}Hc_2 = +\frac{3H}{4}c_2$$

To be noticed how also the translation ξ_5 of the beam is highly influenced, for the same flexibility of the columns, by the ratio of the flexibilities of columns and beams. Setting

$$\xi_5 = \bar{\xi}_5 c$$

with

$$\bar{\xi}_5 = \frac{H}{2}\frac{h^3}{3EI} \qquad c = \frac{2 + 3\kappa}{2 + 12\kappa}$$

one has for c the values of the following table

Fig. 2.71 Diagrams of
internal forces

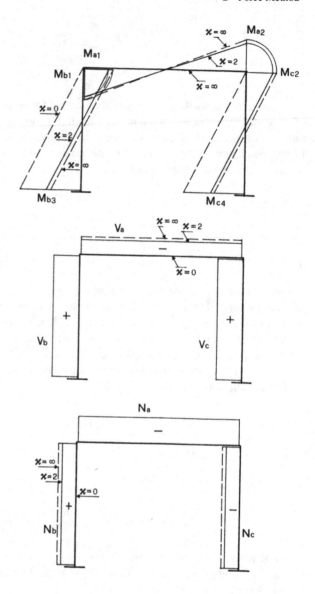

κ	c
∞	1/4
2	4/13
0	1

that indicates, between the two limit situations, a ratio of 1 to 4 of the translation
flexibilities of the portal.

Fig. 2.72 Limit situations of
the column

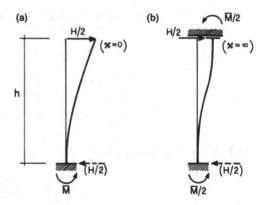

2.4 Examples of Flexibility Analysis

Some examples are proposed first on the calculation of rotations and linear displacements in order to show further applications of the method based on the Mohr's theorem and its corollaries and to complete the set of the "elastic coefficients" of beams to be used in the subsequent applications of the Force Method. Also, if not expressly repeated, the ordinary assumptions of the analysis are confirmed as already presented in the preceding sections.

Exercise No. 1
Calculate the end rotations of the beam of Fig. 2.73a and the deflection of its section 0.

Fig. 2.73 Simply supported
beam with an applied couple

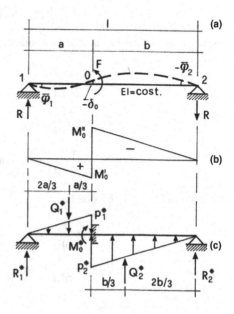

Calculation of the bending moment diagram (see Fig. 2.73b):

$$R = \frac{F}{l}$$

$$M_0' = +Ra = +F\frac{a}{l}$$

$$M_0'' = -Bb = -F\frac{b}{l}$$

Auxiliary beam represented in Fig. 2.73c (see Fig. 2.7b):

$$p_1^* = \frac{|M_0'|}{EI} = +\frac{Fa}{lEI}$$

$$p_2^* = \frac{|M_0''|}{EI} = +\frac{Fb}{lEI}$$

$$Q_1^* = \frac{1}{2}p_1^*a = +\frac{Fa^2}{2lEI}$$

$$Q_2^* = \frac{1}{2}p_2^*a = +\frac{Fb^2}{2lEI}$$

Calculation of the support reactions and moment in section 0

$$R_1^* = Q_1^*\frac{b + a/3}{l} - Q_2^*\frac{2b/3}{l}$$

$$R_1^* = Q_1^*\frac{2a/3}{l} - Q_2^*\frac{a + b/3}{l}$$

$$M_0^* = R_1^*a - Q_1^*a/3$$

Expressions of rotations and displacement:

$$\overline{\varphi}_1 = \frac{+F}{2l^2EI}\left[+\frac{a^3}{3} + a^2b - \frac{2b^3}{3}\right]$$

$$\overline{\varphi}_2 = \frac{+F}{2l^2EI}\left[+\frac{2a^3}{3} - ab^2 - \frac{b^3}{3}\right]$$

$$\delta_0 = \frac{+Fa}{2l^2EI}\left[+\frac{a^3}{3} + a^2b - \frac{2b^3}{3} - \frac{a^2l}{3}\right]$$

With $\alpha = a/l$ and $\beta = b/l$ one has:

$$\overline{\varphi}_1 = +\frac{Fl}{6EI}(+\alpha^3 + 3\alpha^2\beta - 2\beta^3)$$

$$\overline{\varphi}_2 = +\frac{Fl}{6EI}(+2\alpha^3 - 3\alpha\beta^2 - \beta^3)$$

Fig. 2.74 Anti-symmetric case

$$\delta_0 = +\frac{Fl}{6EI}\alpha(+\alpha^3 + 3\alpha^2\beta - 2\beta^3 - \alpha^2)$$

Anti-symmetric case with $\alpha = \beta = 0.5$ (see Fig. 2.74):

$$\overline{\varphi}_1 = +\frac{Fl}{24EI} \quad \text{(clockwise)}$$

$$\overline{\varphi}_2 = -\frac{Fl}{24EI} \quad \text{(clockwise)}$$

$$\delta_0 = 0$$

Case with $\alpha = 1$ and $\beta = 0$ (see Fig. 2.10):

$$\overline{\varphi}_1 = +\frac{Fl}{6EI}$$

$$\overline{\varphi}_2 = +\frac{Fl}{3EI}$$

$$\delta_0 = 0 \quad (= \delta_2)$$

Exercise No. 2

Calculate the end rotations of the beam of Fig. 2.75a and the deflection of its section 0.
Calculation of the bending moment diagram (see Fig. 2.75b):

$$R_1 = P\frac{b}{l}$$

$$M_0 = R_1 a = P\frac{ab}{l}$$

Auxiliary beam represented in Fig. 2.75c (see Fig. 2.7b):

$$p_0^* = \frac{|M_0|}{EI} = \frac{Pab}{lEI}$$

$$Q_1^* = \frac{1}{2}p_0^* a = \frac{Pa^2 b}{2lEI}$$

$$Q_1^* = \frac{1}{2}p_0^* b = \frac{Pab^2}{2lEI}$$

Fig. 2.75 Simply supported
beam with a concentrated
load

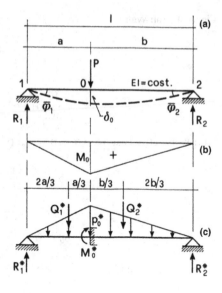

Calculation of the support reactions and moment in section 0:

$$R_1^* = Q_1^* \frac{b + a/3}{1} + Q_2^* \frac{2b/3}{1}$$

$$R_2^* = Q_1^* \frac{2a/3}{1} + Q_2^* \frac{a + b/3}{1}$$

$$M_0^* = R_1^* a - Q_1^* a/3$$

Expressions of rotations and displacement:

$$\overline{\varphi}_1 = \frac{+Pab}{2l^2 EI} \left[+\frac{a^2}{3} + ab - \frac{2b^2}{3} \right]$$

$$\overline{\varphi}_2 = \frac{+Pab}{2l^2 EI} \left[+\frac{2a^2}{3} + ab - \frac{b^2}{3} \right]$$

$$\delta_0 = \frac{+Pa^2 b}{2l^2 EI} \left[+\frac{a^2}{3} + ab + \frac{2b^2}{3} - \frac{al}{3} \right]$$

With $\alpha = a/l$ and $\beta = b/l$, one has:

$$\overline{\varphi}_1 = +\frac{Pl^2}{6EI} \alpha\beta(+\alpha^2 + 3\alpha\beta + 2\beta^2)$$

$$\overline{\varphi}_2 = +\frac{Pl^2}{6EI} \alpha\beta(+2\alpha^2 + 3\alpha\beta + \beta^2)$$

$$\delta_0 = +\frac{Pl^3}{6EI} \alpha^2\beta(+\alpha^2 + 3\alpha\beta + 2\beta^2 - \alpha)$$

Symmetric case with $\alpha = \beta = 0.5$ (see Fig. 2.15):

$$\overline{\varphi}_1 = +\frac{Pl^2}{16EI} \text{ (clockwise)}$$

$$\overline{\varphi}_2 = +\frac{Pl^2}{16EI} \text{ (counter-clockwise)}$$

$$\delta_0 = +\frac{Pl^3}{48EI} \text{ (downwards)}$$

Exercise No. 3

Calculate the end rotations of the beam of Fig. 2.76a and the deflection of the mid-span section 0.

In Fig. 2.77, some notes on the involved geometry are given (drawing of the parabolic curve by tangents, area and centre of its surface parts).

Calculation of the bending moment diagram (see Fig. 2.76b):

$$R = p\frac{l}{2}$$

$$M_0 = R\frac{l}{2} - \frac{pl}{2}\frac{l}{4} = +\frac{pl^2}{8}$$

Fig. 2.76 Simply supported beam with distributed load

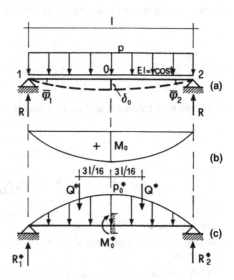

Fig. 2.77 Drawing of the parabolic curve

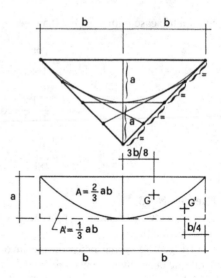

Auxiliary beam represented in Fig. 2.76c (see Fig. 2.7b):

$$p_0^* = \frac{|M_0|}{EI} = \frac{pl^2}{8EI}$$

$$Q^* = \frac{2}{3}\frac{pl^2}{8EI}\frac{1}{2} = \frac{pl^3}{24EI}$$

Calculation of the support reactions and moment in section 0:

$$R_1^* = R_2^* = Q^*$$

$$M_0^* = R_1^*\frac{1}{2} - Q^*\frac{3}{8}\frac{1}{2}$$

Expression of rotations and displacement:

$$\overline{\varphi}_1 = \overline{\varphi}_2 = +\frac{pl^3}{24EI}$$

$$\delta_0 = +\frac{5pl^4}{384EI}$$

Exercise No. 4
Calculate the rotation and deflection of the free end of the cantilever beam of Fig. 2.105a.
   Calculation of the bending moment diagram (see Fig. 2.78b):

$$M_1 = -pl\frac{1}{2} = -\frac{pl^2}{2}$$

Fig. 2.78 Cantilever beam
with distributed load

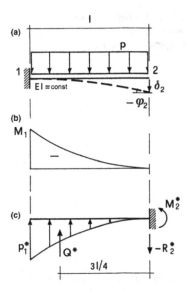

Auxiliary beam represented in Fig. 2.78c (see Fig. 2.7a–c):

$$p_1^* = \frac{|M_1|}{EI} = \frac{pl^2}{2EI}$$

$$Q^* = \frac{1}{3}p_1^* l = \frac{pl^3}{6EI}$$

Calculation of reaction and moment:

$$R_2^* = Q^*$$

$$M_2^* = +Q^* \frac{3l}{4}$$

Expressions of rotation and deflection:

$$\overline{\varphi}_2 = -\frac{pl^3}{6EI} \quad \text{(clockwise)}$$

$$\delta_2 = +\frac{pl^4}{8EI} \quad \text{(downwards)}$$

Exercise No. 5

Calculate the end rotations of the beam of Fig. 2.79a and the deflection of the mid-span section 0.

Fig. 2.79 Simply supported
beam with differential
temperature variation

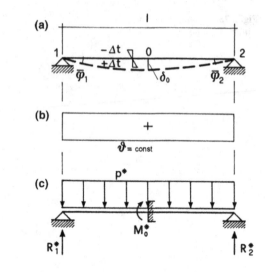

Fig. 2.80 Deformation of
the elementary segment

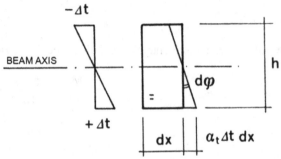

Calculation of the curvature diagram (see Figs. 2.79b and 2.80):

$$\chi = \frac{d\varphi}{dx} = +\frac{2\alpha_t \Delta t}{h} \quad (= \text{const.})$$

Auxiliary beam represented in Fig. 2.79c (see Fig. 2.7b):

$$p^* = |\chi| = \frac{2\alpha_t \Delta t}{h}$$

$$R_1^* = R_2^* = +\frac{p^* l}{2}$$

$$M_0^* = +\frac{p^* l^2}{8}$$

Expressions of rotation and displacement:

$$\overline{\varphi}_1 = \overline{\varphi}_2 = +\frac{\alpha_t l \Delta t}{h}$$

$$\delta_0 = +\frac{\alpha_t l^2 \Delta t}{4h}$$

Exercise No. 6

Calculate the rotations at the supports of beam of Fig. 2.81a and the displacement of the free end 0.

Calculation of bending moment diagram (see Fig. 2.81b):

$$M_2 = -Pa$$

Auxiliary beam represented in Fig. 2.81c (see Figs. 2.7b, c and 2.8a):

$$p_2^* = \frac{|M_2|}{EI} = \frac{Pa}{EI}$$

$$Q_1^* = \frac{1}{2}p_2^* l = \frac{Pal}{2EI}$$

$$Q_2^* = \frac{1}{2}p_2^* a = \frac{Pa^2}{2EI}$$

Fig. 2.81 Simply supported beam with a cantilevering part

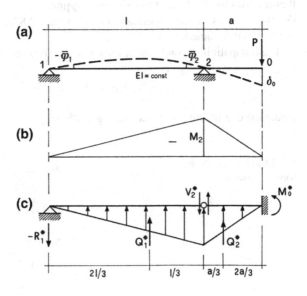

Calculation of reaction in 1, shear in 2 and moment in 0:

$$R_1^* = -Q_1^* \frac{l/3}{l}$$

$$V_2^* = +Q_1^* \frac{2l/3}{l}$$

$$M_0^* = +V_2^* a + Q_2^* 2a/3$$

Expressions of rotations and displacement (with $\alpha = a/l$):

$$\overline{\varphi}_1 = -\frac{1}{6EI} Pa$$

$$\overline{\varphi}_2 = -\frac{1}{3EI} Pa$$

$$\delta_0 = +\frac{Pa^3}{3EI} \left(\frac{1}{\alpha} + 1 \right)$$

2.4.1   Analysis of Fixed-End Beams

In the following exercises, also if not expressly reminded, the same sign conventions previously used are assumed and the standard procedure of the Force Method is applied.

Beam with One Full and One Simple Support
For the analysis of the hyperstatic beam of Fig. 2.82a, the auxiliary structure of Fig. 2.82b is assumed.

The compatibility condition at joint 1 ($\varphi_1 = 0$) is set with:

$$\varphi_{11} X_1 + \varphi_{1o} = 0$$

where the coefficient (evidenced in Fig. 2.25c) is

Fig. 2.82 Beam with one
redundant restrain

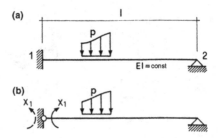

Fig. 2.83 Load conditions

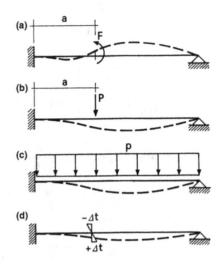

$$\varphi_{11} = \frac{1}{3EI} \quad \left(\varphi_{21} = \frac{1}{6EI}\right)$$

and the solution is obtained with

$$X_1 = -\frac{3EI}{1}\varphi_{1o}$$

For the four load conditions represented in Fig. 2.83a–d, the rotations at the simply supported ends of the auxiliary structure are deduced from the pertinent exercises before presented.

Case "a" with a = 1/2 (see Fig. 2.84)

Fig. 2.84 Beam loaded with a couple at mid-span

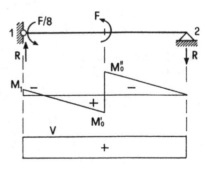

$$\varphi_{10} = +\frac{Fl}{24EI} \qquad\qquad (=-\varphi_{2o})$$

$$X_1 = -\frac{F}{8} = M_1 \qquad\qquad \text{(tensioned the upper fibres)}$$

$$R = \frac{F - X_1}{1} = +\frac{9}{8}\frac{F}{1} = V \qquad \text{(clockwise)}$$

$$M_0' = X_1 + R\frac{1}{2} = +\frac{7}{16}F \qquad \text{(tensioned the lower fibres)}$$

$$M_0'' = M_0' - F = -\frac{9}{16}F \qquad \text{(tensioned the upper fibres)}$$

$$\varphi_2 = \frac{1}{6EI}X_1 - \frac{Fl}{24EI} = -\frac{Fl}{16EI} \text{ (clockwise)}$$

Case "b" with $a = l/2$ (see Fig. 2.85)

$$\varphi_{10} = +\frac{Pl^2}{16EI} \quad (=+\varphi_{2o})$$

$$X_1 = -\frac{3PL}{16} = M_1$$

$$R_1 = +\frac{P}{2} - \frac{X_1}{1} = +\frac{11P}{16} = V_1$$

$$R_2 = +\frac{P}{2} + \frac{X_1}{1} = +\frac{5P}{16} = -V_2$$

$$M_0 = \frac{Pl}{4} + \frac{X_1}{2} = +\frac{5Pl}{32}$$

$$\varphi_2 = \frac{1}{6EI}X_1 + \frac{Pl^2}{16EI} = +\frac{Pl^2}{32EI}$$

Fig. 2.85 Beam with a force
at mid-span

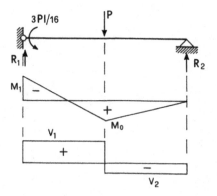

Fig. 2.86 Beam with distributed load

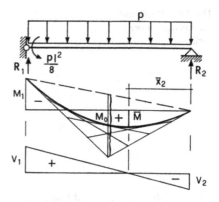

Case "c" (see Fig. 2.86)

$$\varphi_{10} = +\frac{pl^3}{24EI} \quad (= +\varphi_{20})$$

$$X_1 = -\frac{pl^2}{8} = M_1$$

$$R_1 = +\frac{pl}{2} - \frac{X_1}{1} = +\frac{5pl}{8} = V_1$$

$$R_2 = +\frac{pl}{2} + \frac{X_1}{1} = +\frac{3pl}{8} = -V_2$$

$$M_0 = \frac{pl^2}{8} + \frac{X_1}{2} = +\frac{pl^2}{16}$$

$$\bar{x}_2 = -V_2/p = \frac{3l}{8} \quad (= 0.375\,l)$$

$$\overline{M} = R_2\bar{x}_2 - \frac{p\bar{x}_2^2}{2} = +\frac{9pl^2}{128}$$

$$\varphi_2 = \frac{1}{6EI}X_1 + \frac{pl^3}{24EI} = +\frac{Pl^3}{48EI}$$

Case "d" (see Fig. 2.87)

$$\varphi_{1o} = +\frac{\alpha_t l \Delta t}{h} \quad (= +\varphi_{20})$$

$$X_1 = -3EI\frac{\alpha_t \Delta t}{h} = M_1$$

$$R = -\frac{X_1}{1} = +\frac{3EI}{1}\alpha_t \Delta t/h = V$$

$$\varphi_2 = \frac{1}{6EI}X_1 + \frac{\alpha_t l \Delta t}{h} = +\frac{\alpha_t l \Delta t}{2h}$$

Fig. 2.87 Beam with
differential temperature
variation

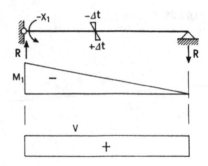

Beam with Two Full End Supports

For the analysis of the hyperstatic beam of Fig. 2.88a, the auxiliary structure of
Fig. 2.88b is assumed.

The compatibility conditions of the end rotations ($\varphi_1 = 0$ and $\varphi_2 = 0$) are set
with:

$$\begin{cases} \varphi_{11}X_1 + \varphi_{13}X_3 + \varphi_{10} = 0 \\ \varphi_{31}X_1 + \varphi_{33}X_3 + \varphi_{30} = 0 \end{cases}$$

The coefficients are (see Fig. 2.31):

$$\varphi_{11} = \varphi_{22} = \frac{1}{3EI}$$
$$\varphi_{12} = \varphi_{21} = \frac{1}{6EI}$$

and the solution is obtained with

$$X_1 = -\frac{2EI}{1}(2\varphi_{1o} - \varphi_{2o})$$
$$X_2 = -\frac{2EI}{1}(2\varphi_{2o} - \varphi_{1o})$$

Fig. 2.88 Beam with two
full end supports

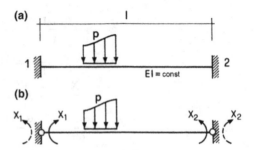

Fig. 2.89 Load conditions

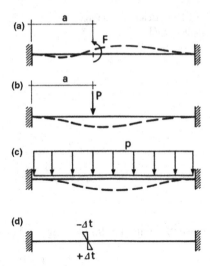

For the four load conditions of Fig. 2.89a–d, the rotations at the end supports of the auxiliary structure are deduced from the same exercises presented at the beginning of the present Section.

Case "a" with a = l/2 (see Fig. 2.90)

$$\varphi_{1o} = -\varphi_{2o} = +\frac{Fl}{24EI}$$

$$X_1 = M_1 = -\frac{F}{4}$$

$$X_2 = M_2 = +\frac{F}{4}$$

$$R = \frac{F - X_1 + X_2}{l} = +\frac{3F}{2l} = V$$

$$M_0' = -M_0'' = X_1 + R\frac{l}{2} = +\frac{F}{2}$$

Fig. 2.90 Beam loaded with a couple at mid-span

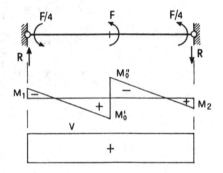

Fig. 2.91 Beam with a force
applied at l/3

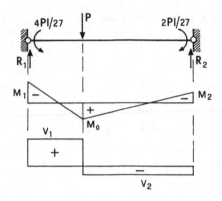

Case "b" with a = l/3 (see Fig. 2.91)

$$\varphi_{1o} = \frac{Pl^2}{6EI}\frac{10}{27} = +\frac{5Pl^2}{81EI}$$

$$\varphi_{2o} = \frac{Pl^2}{6EI}\frac{8}{27} = +\frac{4Pl^2}{81EI}$$

$$X_1 = M_1 = -\frac{4}{27}Pl$$

$$X_2 = M_2 = -\frac{2}{27}Pl$$

$$R_1 = \frac{2P}{3} - \frac{X_1 - X_2}{1} = +\frac{20}{27}P = V_1$$

$$R_2 = \frac{P}{3} + \frac{X_1 - X_2}{1} = +\frac{7}{27}P = -V_2$$

$$M_0 = X_1 + R_1\frac{1}{3} = +\frac{8}{81}Pl$$

Case "c" (see Fig. 2.92)

$$\varphi_{1o} = \varphi_{2o} = +\frac{Pl^3}{24EI}$$

$$X_1 = X_2 = -\frac{Pl^2}{12} = M_1 = M_2$$

$$R_1 = R_2 = +\frac{pl}{2} = V_1 = -V_2$$

$$M_0 = X_1 + R_1\frac{1}{2} - \frac{pl^2}{8} = +\frac{pl^2}{24}$$

Fig. 2.92 Beam with distributed load

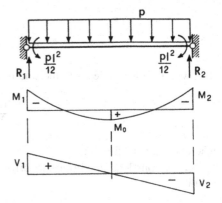

Case "d" (see Fig. 2.93)

$$\varphi_{10} = \varphi_{20} = +\frac{\alpha_t l \Delta t}{h}$$
$$X_1 = X_2 = -2EI\alpha_t \Delta t/h = M$$
$$R_1 = R_2 = 0 = V$$

Beam with an Internal Hinge

For the analysis of the hyperstatic beam of Fig. 2.94a, going without the application of the standard procedure, the auxiliary structure of Fig. 2.94b is assumed. This allows to obtain the solution with one only unknown, as evidenced in the figure. The compatibility condition $\delta_0 = 0$ imposes a null value to the relative displacement between the adjacent ends of the two independent cantilevers in which the beam has been decomposed. Figure 2.94c shows the effect of the hyperstatic force X that, in the case of cantilevers with constant cross section, referring to the example of Fig. 2.17 of Sect. 2.1.3, can be written as

$$\delta_{0X} = \frac{a^3}{3EI_a} + \frac{b^3}{3EI_b}$$

Adding the contribution of the external load, one obtains the equation:

Fig. 2.93 Beam with differential temperature variation

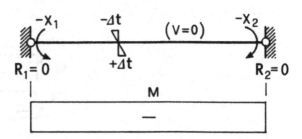

$$\delta_{0x}X + \delta_{00} = 0$$

that, with $I_a = I_b = I$, leads to the solution:

$$X = -\frac{3EI}{l^3}\frac{\delta_{00}}{\alpha^3 + \beta^3}$$

where $\alpha = a/l$ and $\beta = b/l$.

For the four load conditions described in Fig. 2.95, one obtains the following solutions.

Fig. 2.94 Beam with an internal hinge

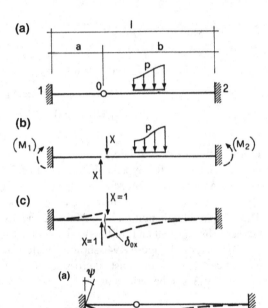

Fig. 2.95 load conditions

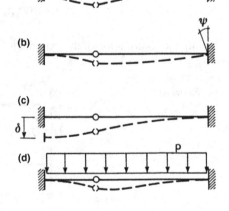

Fig. 2.96 Load effects on
auxiliary structure

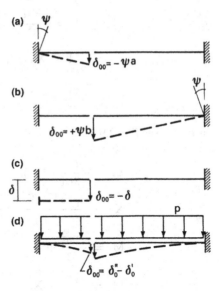

Case "a" (see Fig. 2.96a)

$$\delta_{00} = -\psi a$$

$$X = +\frac{3EI}{l^2}\frac{\alpha}{\alpha^3 + \beta^3}\psi = -V$$

$$M_1 = Xa = +\frac{3EI}{l}\frac{\alpha^2}{\alpha^3 + \beta^3}\psi$$

$$M_2 = -Xb = -\frac{3EI}{l}\frac{\alpha\beta}{\alpha^3 + \beta^3}\psi$$

Case "b" (see Fig. 2.96b):

$$\delta_{00} = +\psi b$$

$$X = -\frac{3EI}{l^2}\frac{\beta}{\alpha^3 + \beta^3}\psi = -V$$

$$M_1 = Xa = -\frac{3EI}{l}\frac{\alpha\beta}{\alpha^3 + \beta^3}\psi$$

$$M_2 = -Xb = +\frac{3EI}{l}\frac{\beta^2}{\alpha^3 + \beta^3}\psi$$

Case "c" (see Fig. 2.96c):

$$\delta_{00} = -\delta$$

$$X = -\frac{3EI}{l^3}\frac{1}{\alpha^3 + \beta^3}\delta = -V$$

$$M_1 = Xa = +\frac{3EI}{l^2}\frac{\alpha}{\alpha^3 + \beta^3}\delta$$

$$M_1 = -Xb = -\frac{3EI}{l^2}\frac{\beta}{\alpha^3 + \beta^3}\delta$$

Case "d" (see Fig. 2.96d):

$$\delta_{00} = -\frac{pl^4}{8EI}(\alpha^4 - \beta^4)$$

$$X = +\frac{3pl}{8}\frac{\alpha^4 - \beta^4}{\alpha^3 + \beta^3} = -V_0$$

$$M_1 = -\frac{pa^2}{2} + Xa = -\frac{pl^2}{8}\left[4\alpha^2 - 3\alpha\frac{\alpha^4 - \beta^4}{\alpha^3 - \beta^3}\right]$$

$$M_2 = -\frac{pb^2}{2} - Xb = -\frac{pl^2}{8}\left[4\beta^2 - 3\beta\frac{\alpha^4 - \beta^4}{\alpha^3 - \beta^3}\right]$$

$$V_1 = +pa - X = +\frac{pl}{8}\left[8\alpha - 3\frac{\alpha^4 - \beta^4}{\alpha^3 - \beta^3}\right]$$

$$V_2 = -pb - X = -\frac{pl}{8}\left[8\beta - 3\frac{\alpha^4 - \beta^4}{\alpha^3 - \beta^3}\right]$$

For the first three cases, the diagrams of bending moment has a linear variation between the values M_1 and M_2 passing for zero on the hinge and diagrams of shear force with a constant value V. For the last case, the diagrams are shown in Fig. 2.97, where in particular it is assumed $\alpha = 1/3$ and $\beta = 2/3$ obtaining the values:

$$X = -\frac{5}{24}pl = -V_0$$

Fig. 2.97 Diagrams of moment and shear for case "d"

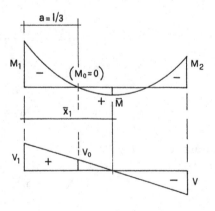

$$M_1 = -\frac{pl^2}{8}$$

$$M_2 = -\frac{pl^2}{12}$$

$$V_1 = +\frac{13}{24}pl$$

$$V_2 = -\frac{11}{24}pl$$

$$\bar{x}_1 = V_1/p = \frac{13}{24}l$$

$$\overline{M} = M_1 + V_1\bar{x}_1 - \frac{p\bar{x}_1^2}{2} = +0.17\frac{pl^2}{8}$$

2.4.2 Analysis of Continuous Beams

The following exercises, besides giving some simple examples of application of the standard procedure, are aimed to show the influence, on the behaviour of continuous beams, of some structural parameters, such as the ratio between the length of the spans, the different load arrangements or the effectiveness of the end restraints of the beam. Indications will provided on the response types, on their variability and on the possible different stress distributions so to make friendly the analysis of this type of structures when referred to real design applications.

Continuous Beam with Two Spans
For the analysis of the continuous beam of Fig. 2.98a, the auxiliary structure of Fig. 2.98b is assumed. The compatibility condition

$$\varphi_{22}X_2 + \varphi_{20} = 0$$

Fig. 2.98 Continuous beam with two spans

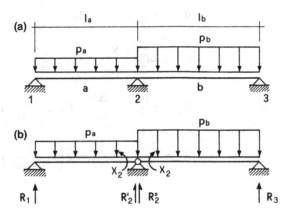

Fig. 2.99 Curve of $\kappa = \kappa(\kappa_0)$

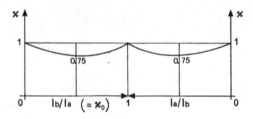

is written as

$$\left(\frac{l_a}{3EI_a} + \frac{l_b}{3EI_b}\right)X_2 + \left(\frac{pl_a^3}{24EI_a} + \frac{pl_b^3}{24EI_b}\right) = 0$$

Case "1" with $p_a = p_b = p$, $l_a = l$ and $l_b = \kappa_0 l$ ($I_a = I_b$)
    The solution is:

$$X_2 = -\frac{pl^2}{8}\frac{1+\kappa_0^3}{1+\kappa_0} = -\overline{M}\kappa$$

with $\overline{M} = pl^2/8$. Figure 2.99 gives the curve of the value of κ as a function of the ratio $\kappa_0 = l_b/l_a$ between the lengths of the two spans. The point $\kappa_0 = 0$ corresponds to a full support at joint 2 without the second span and leads to the same solution of the beam of Fig. 2.86. To this same solution leads also the symmetric arrangement with $\kappa_0 = 1$, while the values for $\kappa_0 > 1$ can be deduced from the ones for $\kappa_0 < 1$ simply exchanging the reference length l ($l_b = l$ with $l_a = \kappa_0 l$).

Figure 2.100 shows the diagrams of the bending moment for some values of the ratio κ_0. One can see that there are small differences in the behaviour of the longer span while in the shorter span, the diagram can also overturn. Similar big differences there are for the shear force in the shorter span as can be noticed in Fig. 2.101.

For completeness sake, the calculations of the reactions are set here under, from which one can directly deduce the shear values of the correspondent sections:

$$R_1 = +\frac{pl}{2} + \frac{X_2}{l} = \overline{R}\left(1 - \frac{\kappa}{4}\right) \qquad (=+V_{a1})$$

$$R_2' = +\frac{pl}{2} - \frac{X_2}{l} = \overline{R}\left(1 + \frac{\kappa}{4}\right) \qquad (=-V_{a2})$$

$$R_2'' = +\frac{pl}{2}\kappa_0 - \frac{X_2}{l\kappa_0} = \overline{R}\left(\kappa_0 + \frac{\kappa}{4\kappa_0}\right) \quad (=+V_{b2})$$

$$R_3 = +\frac{pl}{2}\kappa_0 + \frac{X_2}{l\kappa_0} = \overline{R}\left(\kappa_0 - \frac{\kappa}{4\kappa_0}\right) \quad (=-V_{b3})$$

Fig. 2.100 Bending moment diagrams for different κ_o ratios

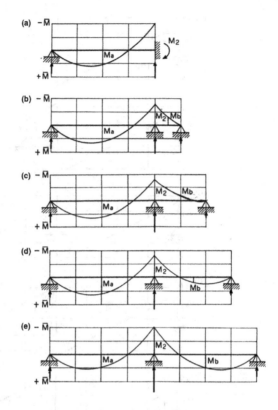

where $\overline{R} = pl/2$. The bending moments at middle of the two spans are eventually evaluated with

$$M_a = +\frac{pl^2}{8} + \frac{X_2}{2} = \overline{M}\left(1 - \frac{\kappa}{2}\right)$$

$$M_b = +\frac{pl^2}{8}\kappa_o^2 + \frac{X_2}{2} = \overline{M}\left(\kappa_o^2 - \frac{\kappa}{2}\right)$$

Case "2" with $p_a = p$, $p_b = 0$, $l_a = l$ and $l_b = \kappa_o l$ ($I_a = I_b$)

The solution is:

$$X_2 = -\frac{pl^2}{8}\frac{1}{1+\kappa_o} = -\overline{M}\kappa_a$$

for which Fig. 2.102 gives the curve of variation of κ_a as a function of the ratio κ_o between the length of the spans. In this case, the hyperstatic moment X_2 is ever decreasing for longer lengths of the second not loaded span as it becomes ever more flexible.

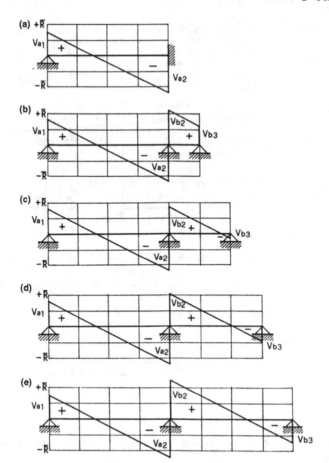

Fig. 2.101 Shear force diagrams for different κ_o ratios

Fig. 2.102 Curve of $\kappa_a = \kappa_a(\kappa_o)$

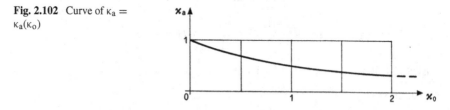

Figures 2.103 and 2.104 show the diagrams of the bending moment and shear force for some values of κ_o. At the limit $\kappa_o \to \infty$, the situation of the loaded span tends to that of the simply supported beam with $X_2 = 0$ and $M_a = pl^2/8$.

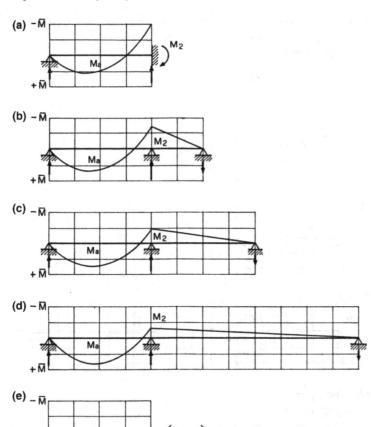

Fig. 2.103 Bending moment diagrams for different κ_o ratios

For completeness sake, the calculations of the reactions are set here under, from which one can directly deduce the shear values at the ends of the spans:

$$R_1 = \overline{R}\left(1 - \frac{\kappa_a}{4}\right) \quad (=+V_{a1})$$

$$R'_2 = \overline{R}\left(1 + \frac{\kappa_a}{4}\right) \quad (=-V_{a2})$$

$$R''_2 = \overline{R}\left(\frac{\kappa_a}{4\kappa_o}\right) \quad (=+V_{b2})$$

$$R_3 = -\overline{R}\left(\frac{\kappa_a}{4\kappa_o}\right) \quad (=-V_{b3})$$

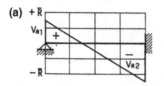

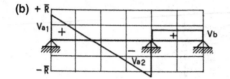

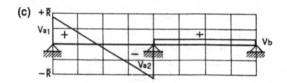

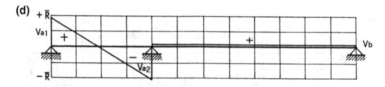

Fig. 2.104 Shear force diagrams for different κ_o ratios

where $\overline{R} = pl/2$. The bending moment at the middle of the first span is:

$$M_a = +\frac{pl^2}{8} + \frac{X_2}{2} = \overline{M}\left(1 - \frac{\kappa_a}{2}\right)$$

Continuous Beam with Three Spans

For the analysis of the continuous beam of Fig. 2.105a, the auxiliary structure of Fig. 2.105b is assumed. The set of compatibility conditions

$$\begin{cases} \varphi_{22}X_2 + \varphi_{23}X_3 + \varphi_{20} = 0 \\ \varphi_{32}X_2 + \varphi_{33}X_3 + \varphi_{30} = 0 \end{cases}$$

is written as:

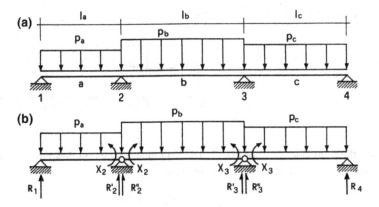

Fig. 2.105 Continuous beam with three spans

$$\begin{cases} \left(\dfrac{l_a}{3EI_a} + \dfrac{l_b}{3EI_b}\right)X_2 + \dfrac{l_b}{6EI_b}X_3 + \left(\dfrac{p_a l_a^3}{24EI_a} + \dfrac{p_b l_b^3}{24EI_b}\right) = 0 \\[3mm] \dfrac{l_b}{6EI_b}X_2 + \left(\dfrac{l_b}{3EI_b} + \dfrac{l_c}{3EI_c}\right)X_3 + \left(\dfrac{p_b l_b^3}{24EI_b} + \dfrac{p_c l_c^3}{24EI_c}\right) = 0 \end{cases}$$

For a symmetric beam with $I_a = I_b = I_c = I$, $l_b = l$ and $l_a = l_c = \kappa_o l$, one has

$$\begin{cases} 2(\kappa_o + 1)X_2 + X_3 = -\dfrac{l^2}{4}(\kappa_o^3 p_a + p_b) \\[3mm] X_2 + 2(\kappa_o + 1)X_3 = -\dfrac{l^2}{4}(p_b + \kappa_o^3 p_c) \end{cases}$$

that, for a symmetric load ($p_a = p_c$), leads, with $X_2 = X_3$, to one only equation:

$$(2\kappa_o + 3)X_2 = -\frac{l^2}{4}(\kappa_o^3 p_a + p_b)$$

Case "1" with $p_a = p_b = p_c = p$ over all the structure

The solution is:

$$X_2 = X_3 = -\frac{pl^2}{8} 2 \frac{1 + \kappa_o^3}{3 + 2\kappa_o} = -\overline{M}\kappa$$

with $\overline{M} = pl^2/8$. Figure 2.106 shows the curve that, similar to that of Fig. 2.99, provides the values of κ as a function of the ratio κ_o between the span lengths. The point $\kappa_o = 0$ corresponds to the arrangement with two full end supports without the lateral spans and leads to the same solution of the beam of Fig. 2.92. For $l_a = l_c > l_b$, it is more meaningful to assume as reference the length of the lateral spans,

Fig. 2.106 Curve of $\kappa =$
$\kappa(\kappa_0)$

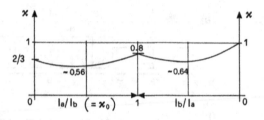

following the right side curve of Fig. 2.106 that tends, as the middle span decreases, to the value $\kappa = 1$ corresponding to the hyperstatic moment

$$X_2 = -\overline{M}\kappa = -\frac{pl^2}{8} \quad (= X_3)$$

where in $\overline{M} = pl^2/8$ the longer span length is introduced.

Figures 2.107 and 2.108 show the diagrams of bending moment and shear force for some values of the ratio κ_o between the span lengths. On these diagrams, one can point out again the relevant influence of the longer spans on the shorter contiguous ones.

For completeness sake, the calculations of the reactions are set here under:

$$R_1 = +\frac{pl}{2}\kappa_o + \frac{X_2}{l\kappa_o} = +\overline{R}\left(\kappa_o - \frac{\kappa}{4\kappa_o}\right) = R_4$$

$$R_2' = +\frac{pl}{2}\kappa_o - \frac{X_2}{l\kappa_o} = +\overline{R}\left(\kappa_o + \frac{\kappa}{4\kappa_o}\right) = R_3''$$

$$R_2'' = +\frac{pl}{2} = \overline{R} = R_3'$$

The bending moments at mid-spans are:

$$M_a = +\frac{pl^2}{8}\kappa_o^2 + \frac{X_2}{2} = \overline{M}\left(\kappa_o^2 - \frac{\kappa}{2}\right) = M_c$$

$$M_b = +\frac{pl^2}{8} + X_2 = \overline{M}(1 - \kappa)$$

The limits $\pm M$ and $\pm V$ marked in Figs. 2.107 and 2.108 refer to the values calculated with the longer span length.

Case "2" with $p_a = p_c = 0$ and $p_b = p$ on symmetric structure
The solution is:

$$X_2 = X_3 = -\frac{pl^2}{8}\frac{2}{3 + 2\kappa_o} = -\overline{M}\kappa_b$$

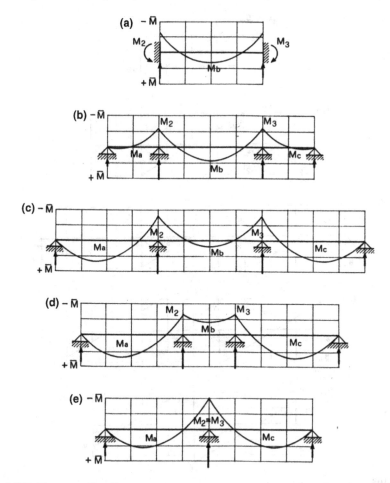

Fig. 2.107 Diagrams of bending moment

for which Fig. 2.109 gives the variation curve of κ_b as a function of the ratio between the length of the lateral spans and the one of the central loaded spans. As their length increases, the lateral not loaded spans become more and more flexible. By consequence also, the value of the hyperstatic moments decreases up to reach value zero at the limit $\kappa_o \to \infty$ for which the central span assumes the arrangement of a simply supported beam.

Figures 2.110 and 2.111 show the diagrams of bending moment and shear force for some values of the ratio κ_o between the span lengths.

For completeness sake, the calculations of the reactions are set here under:

$$R_1 = -\overline{R}\frac{\kappa_b}{4\kappa_o} = R_4$$

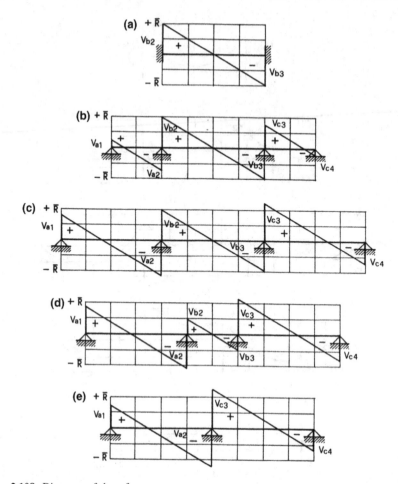

Fig. 2.108 Diagrams of shear force

Fig. 2.109 Curve of $\kappa_b = \kappa_b(\kappa_0)$

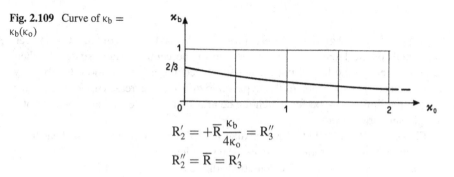

$$R_2' = +\overline{R}\frac{\kappa_b}{4\kappa_0} = R_3''$$

$$R_2'' = \overline{R} = R_3'$$

The moment at the mid-section of the central span is:

$$M_b = +\overline{M}(1 - \kappa_b)$$

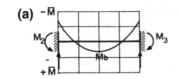

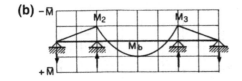

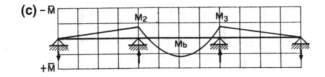

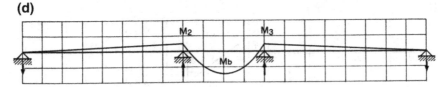

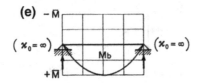

Fig. 2.110 Diagrams of bending moments

Case "3" with $p_a = p$, $p_b = p_c = 0$ and $\kappa_0 = 1$

From the compatibility equations

$$\begin{cases} 4X_2 + X_3 = -\dfrac{pl}{4} \\[2mm] X_2 + 4X_3 = 0 \end{cases}$$

one obtains the solution:

$$X_2 = -\frac{pl^2}{8}\frac{8}{15} = -\overline{M}\frac{8}{15}$$

$$X_3 = +\frac{pl^2}{8}\frac{2}{15} = +\overline{M}\frac{2}{15}$$

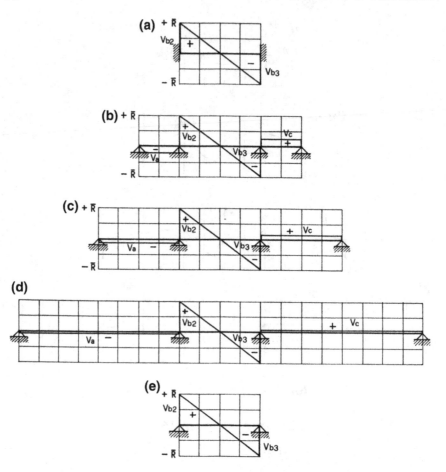

Fig. 2.111 Diagrams of shear force

that leads to the diagrams of Fig. 2.112. Compared to those of the corresponding case of beam with two spans (see Figs. 2.103c and 2.104c), these diagrams show the small influence of the distant third span on the first loaded one.

The analysis is completed with the calculation of the reactions and of the moment at the middle of the loaded span:

$$R_1 = +\frac{pl}{2} + \frac{X_2}{1} = +\bar{R}\frac{13}{15}$$

$$R_2' = +\frac{pl}{2} - \frac{X_2}{1} = +\bar{R}\frac{17}{15}$$

$$R_2'' = -R_3' = -\frac{X_2 - X_3}{1} = \bar{R}\frac{1}{6}$$

$$R_3'' = -R_4 = -\frac{X_3}{1} = \bar{R}\frac{1}{30}$$

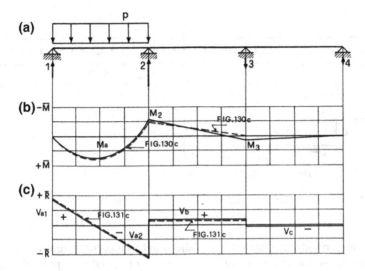

Fig. 2.112 Continuous beam with three spans

$$M_a = +\frac{pl^2}{8} + \frac{X_2}{2} = \overline{M}\frac{11}{15}$$

where again it is set $\overline{R} = pl/2$ and $\overline{M} = pl^2/8$.

2.4.3 Multi-span Continuous Beams

The behaviour of multi-span continuous beams is now examined. To this end, let's consider the beam of Fig. 2.113a that, to simplify the analysis, is assumed symmetric, with constant cross section and with spans all of the same length. One can refer to the auxiliary halved structure of Fig. 2.113b, that is analysed with the only two unknowns X_2 and X_3 with the compatibility equations:

$$\begin{cases} \varphi_{22}X_2 + \varphi_{23}X_3 + \varphi_{20} = 0 \\ \varphi_{32}X_2 + \varphi_{33}X_3 + \varphi_{30} = 0 \end{cases}$$

Case "1" with load distributed over all the continuous beam
For a constant load as shown in Fig. 2.113, one has:

$$\begin{cases} \left(\dfrac{1}{3EI} + \dfrac{1}{3EI}\right)X_2 + \dfrac{1}{6EI}X_3 + \left(\dfrac{pl^3}{24EI} + \dfrac{pl^3}{24EI}\right) = 0 \\ \dfrac{1}{6EI}X_2 + \left(\dfrac{1}{3EI} + \dfrac{1}{4EI}\right)X_3 + \left(\dfrac{pl^3}{24EI} + \dfrac{pl^3}{48EI}\right) = 0 \end{cases}$$

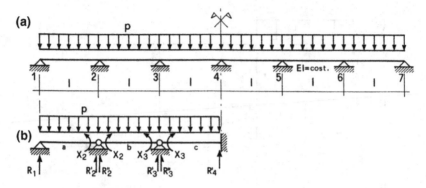

Fig. 2.113 Multi-span continuous beam

written using, besides the flexibilities of the simply supported beam, also the flexibility and the pertinent rotation of the beam with a full and a simple end supports (see Figs. 2.26 and 2.86).

After the opportune simplifications, one obtains:

$$
\begin{cases}
8X_2 + 2X_3 = -\dfrac{4}{4}pl^2 \\[2mm]
2X_2 + 7X_3 = -\dfrac{3}{4}pl^2
\end{cases}
$$

from which the solution is calculated as:

$$
X_2 = -\frac{pl^2}{8}\frac{11}{13} \quad (= M_2)
$$

$$
X_3 = -\frac{pl^2}{8}\frac{8}{13} \quad (= M_3)
$$

$$
M_4 = -\frac{pl^2}{8} - \frac{X_2}{2} = -\frac{pl^2}{8}\frac{9}{13}
$$

$$
M_a = +\frac{pl^2}{8} + \frac{X_2}{2} = +\frac{pl^2}{8}\frac{15}{26}
$$

$$
M_b = +\frac{pl^2}{8} + \frac{X_2 + X_3}{2} = +\frac{pl^2}{8}\frac{7}{26}
$$

$$
M_c = +\frac{pl^2}{8} + \frac{X_3 + M_4}{2} = +\frac{pl^2}{8}\frac{9}{26}
$$

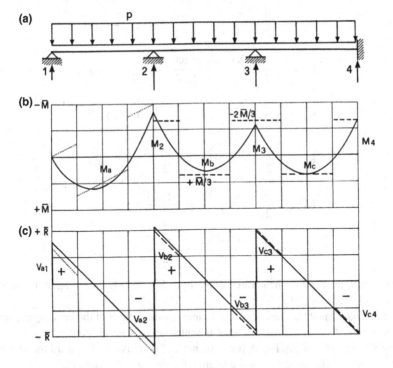

Fig. 2.114 Diagrams of internal forces

In Fig. 2.114b, the diagram of the bending moment is indicated, having marked the limits:

$$\pm\overline{M} = \pm\frac{pl^2}{8}$$

For the internal spans, also the limits $-pl^2/12 = -2\overline{M}/3$ and $+pl^2/24 = +\overline{M}/3$ are indicated by dashed lines with reference to the behaviour of the beam with two full end supports.

It can be noticed that, for the internal spans, the bending moment diagram gets closer to the one of the beams with two full end supports, while for the lateral span, the diagram is close to the one of the beams with one full and one simple end supports. The trend of this latter diagram is indicated by the dotted lines of Fig. 2.114b.

The calculations of the reactions are eventually set as:

$$R_1 = +\frac{pl}{2} + \frac{X_2}{1} = +\frac{pl}{2}\frac{41}{52} \qquad (= +V_{a1})$$

$$R_2' = +\frac{pl}{2} - \frac{X_2}{1} = +\frac{pl}{2}\frac{63}{52} \qquad (= -V_{a2})$$

$$R_2'' = +\frac{pl}{2} - \frac{X_2 - X_3}{1} = +\frac{pl}{2}\frac{55}{52} \quad (= +V_{b2})$$

$$R_3' = +\frac{pl}{2} + \frac{X_2 - X_3}{1} = +\frac{pl}{2}\frac{49}{52} \quad (= -V_{b3})$$

$$R_3'' = +\frac{pl}{2} - \frac{X_3 - X_4}{1} = +\frac{pl}{2}\frac{51}{52} \quad (= +V_{c3})$$

$$R_4' = +\frac{pl}{2} + \frac{X_3 - X_4}{1} = +\frac{pl}{2}\frac{53}{52} \quad (= -V_{c4})$$

These values lead to the shear force diagram of Fig. 2.114c where again the limits $\pm\overline{R} = \pm pl/2$ are indicated.

On the diagram of the shear force, the same considerations of the bending moment can be repeated about the behaviour of the internal and lateral spans.

From the results above listed, it can be noticed that, assuming for all the internal "current" spans, the moments of the beam with two full end supports:

$$M_i = -\frac{pl^2}{12} = -0.083pl^2 = -0.667\overline{M} \text{ over the supports}$$

$$M_o = +\frac{pl^2}{24} = +0.042pl^2 = +0.333\overline{M} \text{ at the mid-spans,}$$

one makes mistakes of underestimation that don't reach the 3% of $\overline{M}$ and mistakes of overestimation that don't reach the 7% of the same reference term $\overline{M} = pl^2/8$.

For the lateral spans, the solution above gives increased values, with

$$M_i' = -\frac{pl^2}{9.5} = -0.106pl^2 = -0.846\overline{M}$$

over the internal support; this leads to a maximum positive moment $\overline{M}(1 - c/4)^2$ in the span that, with $c = 0.846$, is set as:

$$M_o' = +\frac{pl^2}{12.9} = +0.078pl^2 = +0.622\overline{M}$$

Case "2" with only some spans loaded

Besides the heavier burden deriving from the loss of symmetry that imposes to operate with a set of six equations, for partial loads, there is no problem in the setting of the compatibility conditions following the general procedure presented at Sect. 2.2.2 (see Figs. 2.38 and 2.39). But in this session, an approximate method of analysis is presented that follows the comments made on the results of the preceding applications.

With reference to the continuous beam with three spans of Fig. 2.112, it has been pointed out as the behaviour of the lateral loaded span was little influenced by the presence of the more distant span: The "exact" diagrams drawn with continuous line are little different from those drawn with dashed line obtained from a partial scheme with only two spans deleting the third more distant span.

On the not loaded spans, the differences are larger, but they are still modest with reference to the magnitude of the maximum internal forces that, for these spans, are mainly produced by the loads directly applied on them. This is stated in consideration of the real structural situations that are composed by the combination first of the permanent loads related to the weight of the construction elements extended to all

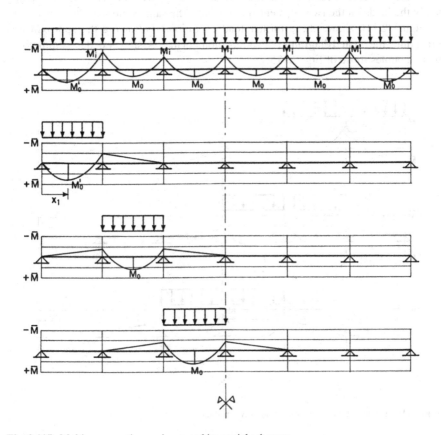

Fig. 2.115 Multi-span continuous beam and its partial schemes

the structure and then of the different possible arrangements of the service loads applied to the single spans of the same structure.

A similar discourse can be repeated for loads distributed only on one span. The corresponding diagrams obtained on the partial scheme, made of the loaded span plus the adjacent spans at its two sides, are little different from the "exact" ones that take into account also the more distant spans.

So for the given continuous beam of Fig. 2.113a, the effects of the partial load conditions can be evaluated in an approximate way using the partial structural scheme mentioned above. Figure 2.115 describes this criterion, where the corresponding diagrams, referred to the bending moment, are taken from Figs. 2.103c and 2.110c. The upper diagram of Fig. 2.115 refers to the global load, uniformed on the internal spans with the approximate "current" behaviour described for the *Case* "1" already discussed.

For the variable loads with free arrangement, the condition that involves one only span is the one that leads to the maximum positive value of the internal bending moment in the same span, unless minor contributions coming from distant spans. Instead, in order to obtain the maximum negative value over one support, one should apply the loads on the two adjacent spans around the same support.

Figure 2.116 shows these further load conditions. Following the approximate procedure, the related diagrams of bending moment may be simply deduced summing up by couples those of Fig. 2.115.

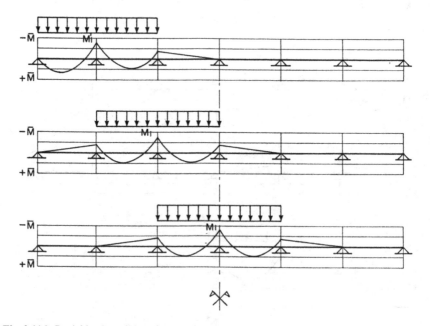

Fig. 2.116 Partial load conditions for negative moments

For the case of concern, the maximum negative and positive moments due to the action of the variable loads can be expressed by the following equations:

$$M_i = -\frac{pl^2}{10.0} = -0.100pl^2 = -0.800\overline{M}$$

$$M_o = +\frac{pl^2}{13.3} = +0.075pl^2 = +0.600\overline{M}$$

$$M_i' = -\frac{pl^2}{8.9} = -0.112pl^2 = -0.896\overline{M}$$

$$M_i = +\frac{pl^2}{10.4} = +0.096pl^2 = +0.768\overline{M}$$

that come from the results related to the continuous beam with three spans analysed at Sect. 2.4.2 and are valid, in an approximate way, for spans all of the same length.

Chapter 3
Displacement Method

Abstract In this chapter, renouncing a more general presentation and focusing on the operational aspects, the Displacement Method is interpreted as a derivation of the Force Method applying in a "dual" way the compatibility and equilibrium conditions with a simple exchange of the static and geometric quantities. In this way, the standard procedure is presented, based on the use of the stiffness parameters of the beams for the formulation of the equilibrium conditions that govern the problem. In its general application, this procedure represents the most effective algorithm for the frame analysis to which the codes for computer-automated calculation conform. A thorough set of examples is presented with reference to non-sway and sway plane and spatial frames.

3.1 General Formulation of the Method

As the Force Method employs, for the formulation of the *compatibility equations*, the elastic coefficients given by the beam flexibilities and end rotations due to loads, the Displacement Method presented in this chapter employs, in its *equilibrium equations*, new elastic coefficients that represent the beam *stiffness* and the full support *end forces*.

The stiffness is a parameter related to the deformation behaviour of beams and represents the forces and moments that display at their ends when unity displacements or rotations are applied. This definition, that is the "dual" of that of the flexibilities, is referred to the arrangement of the single beam belonging to the typical auxiliary structure of the Displacement Method. As specified more in detail below, this arrangement consists of the beam with two full end supports, while in a "dual" way the simply supported beam is the typical arrangement of the standard procedure of the Force Method.

Figure 3.1 shows the quantities of concern, that is:

- the *direct rotation stiffness of the first end* k_1;
- the *indirect rotation stiffness* k_i;
- the *direct rotation stiffness of the second end* k_2;
- the *indirect translation stiffness of the first end* k_{v1};

© Springer Nature Switzerland AG 2019

G. Toniolo, *Introduction to Frame Analysis*, Springer Tracts
in Civil Engineering, https://doi.org/10.1007/978-3-030-14664-1_3

– the *indirect translation stiffness of the second end* k_{v2};
– the *direct translation stiffness* k_v.

The rotation stiffness expresses moments due to unity rotations and therefore is measured in N m; the indirect translation stiffness expresses shear forces due to unity rotations or moments due to unity translations and therefore is measured in N; the direct translation stiffness expresses shear force due to translation and therefore is measured in N/m. In Fig. 3.1, the equalities of the indirect effects are also pointed out as deducible from the pertinent reciprocity theorems.

For the action of loads, there are the full support end forces and moments such as those shown in Fig. 3.2.

The calculation of the stiffness and end forces due to loads is performed through the analysis of the hyperstatic beam to which they refer. This analysis has already been elaborated in the preceding chapter. Referring, for example, to the solution obtained for the beam with constant cross section (see Figs. 2.34 and 2.35), one has the following rotation stiffness:

$$k_1 = k_2 = \frac{4EI}{l}$$

Fig. 3.1 Representation of beam stiffness

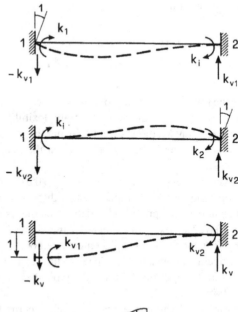

Fig. 3.2 Full support end forces and moments

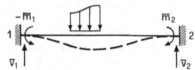

$$k_i = \frac{2EI}{1}$$

$$k_{v1} = k_{v2} = \frac{6EI}{1^2}$$

$$k_v = \frac{12EI}{1^3}$$

For beams with variable cross section, one has to elaborate the equations:

$$k_1 = \frac{d_2}{d_1 d_2 - d_i^2}$$

$$k_2 = \frac{d_1}{d_1 d_2 - d_i^2}$$

$$k_i = \frac{d_i}{d_1 d_2 - d_i^2}$$

that express the solution of the compatibility equation system passing through the numerical evaluation of the flexibilities d_1, d_2, d_i with the algorithms described in Sect. 2.1.2. The translation stiffness is then deduced by the simple rotation equilibria of the beam:

$$k_{v1} = \frac{k_1 + k_i}{1}$$

$$k_{v2} = \frac{k_2 + k_i}{1}$$

$$k_v = \frac{k_{v1} + k_{v2}}{1}$$

One can notice how all is based on a "few recurrent expressions" compatible with an effective standardization of the procedures. One can also widen the set of tabled coefficients with the inclusion of the case of beam with one full and one simple end supports, so to save unknowns as made for the Force Method. With reference to the beam of Fig. 2.27, one has the stiffness:

$$k_1 = \tfrac{3EI}{1} \qquad (= 1/d_1)$$
$$k_2 = k_i = 0$$
$$k_{v1} = \tfrac{3EI}{1^2} \qquad (= k_1/1)$$
$$k_{v2} = 0$$
$$k_v = \tfrac{3EI}{1^3} \qquad (= k_{v1}/1)$$

In order to analyse, in addition to the flexural behaviour of the bending moment and shear force, the axial and torsional ones, the following quantities are introduced:

– the *axial stiffness* k_x which is the axial force corresponding to a unity elongation of the beam;

– the *torsional stiffness* k_t which is the torsional moment corresponding to a unity rotation between the end sections of the beam.

Through the solution of the simple hyperstatic problems of Fig. 3.3 and on the basis of the elastic coefficients d_x and d_t defined in Sect. 2.3, one has:

$$k_x = \frac{EA}{l} \ (= 1/d_x)$$

$$k_t = \frac{GJ}{l} \ (= 1/d_t)$$

To these quantities, the load effects have to be added with the corresponding end forces and moments as shown in Fig. 3.4.

The sign conventions introduced in Sect. 2.1 are eventually reminded (see Fig. 2.1). Dealing with frame analyses, *reference systems* of axis will be assumed as indicated in Fig. 3.5: a global one X, Y, Z to which the whole structural assembly is referred; plus the local systems x, y, z connected to any single member.

For a plane problem, usually the X-axis is set horizontally with rightwards orientation, the Y-axis is set vertically with upwards orientation and the Z-axis is oriented towards who is looking. All the local systems have the z-axis parallel to the homonym one of the global system, while the orientation of the other two axes is defined by the angle α as shown in Fig. 3.5.

Fig. 3.3 Definition of axial and torsional stiffness

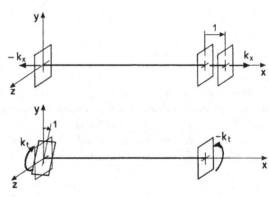

Fig. 3.4 Effects of axial and torsional actions

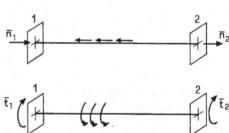

Fig. 3.5 Reference systems
of a plane frame

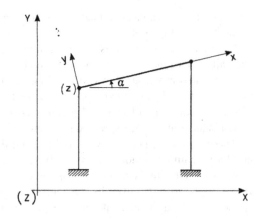

For the final analysis of any single member, after the solution of the overall set of equilibrium conditions and in view of the drawing of the diagrams of internal forces, specific additional conventions may be set.

3.1.1 The Dual Approach

In this section, renouncing a more general presentation and focusing on the operational aspects, the Displacement Method is interpreted as a derivation of the Force Method applying in a "dual" way the compatibility and equilibrium conditions with a simple exchange of the static and geometric quantities.

So, let's consider, for a clearer presentation, the continuous beam of Fig. 3.6a, the same that was analysed at Sect. 2.2.2 with the Force Method (see Fig. 2.40).

From the given one, an auxiliary structure shall be deduced with the addition of supports that fix any free displacement degree at the joints: in the case of concern,

Fig. 3.6 Given (**a**) and
auxiliary (**b**) beam

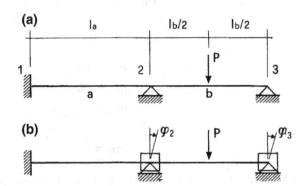

the rotation constraints are added as indicated by the "wedges" placed on the joints of the beam of Fig. 3.6b.

Such an auxiliary beam is called "geometrically determined" in analogy to the statically determined auxiliary beam of the Force Method.

In the auxiliary structure with fixed joints, the members are reciprocally independent and can be analysed separately so that a standard procedure can be applied where the only type of involved member is the beam with two end full supports which behaviour can be expressed by a few recurrent formulae.

The equivalence between the auxiliary and the given beam can be set if on the auxiliary beam one evidences expressly, as geometrical actions, the displacement freedoms fixed with the additional supports. So, again with reference to the beam of Fig. 3.6b, besides the given load P, one shall apply the rotations φ_1 and φ_2 correspondent to the displacement freedoms present in the given beam.

These rotations are the *geometrical unknowns* of the problem following the method under presentation, and they already fulfil the compatibility conditions since they do not break the continuity of the elastic deformed profile of the beam.

Among all the possible values of these unknowns, one has to define eventually the ones that fulfil the other equivalence conditions, that is the self-equilibrium of the joints of the given structure. Therefore on the auxiliary structure, the reactions of the additional restraints shall be put equal to zero.

In formulating these *equilibrium conditions*

$$\begin{cases} m_2(P; \varphi_2, \varphi_3) = 0 \\ m_3(P; \varphi_2, \varphi_3) = 0 \end{cases}$$

within the linear elastic behaviour considered, the superposition and proportionality of effects are applied as described in Fig. 3.7: that is, for each of the additional restraints of the auxiliary beam, the moment contributions due to external load and to the single geometric unknowns are summed up, arriving to the equations:

$$\begin{cases} m_{20} + m_{22}\varphi_2 + m_{23}\varphi_3 = 0 \\ m_{30} + m_{32}\varphi_2 + m_{33}\varphi_3 = 0 \end{cases}$$

with

$$m_{20} = \overline{m}_{b1}$$
$$m_{30} = \overline{m}_{b2}$$
$$m_{22} = k_{a2} + k_{b1}$$
$$m_{32} = k_{bi} = m_{23}$$
$$m_{33} = k_{b2}$$

where the clockwise moments rendered by the supports to the beam ends are assumed positive.

Fig. 3.7 Superimposition of effects

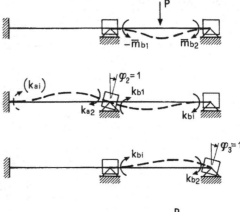

Fig. 3.8 Equivalent isostatic beam

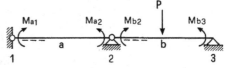

Once the set of equilibrium equations is solved, one can deduce the moments at the ends of all the members with the same superposition and proportionality of effects applied to the auxiliary structure (see again Fig. 3.7):

$$M_{a1} = +k_{ai}\varphi_2$$
$$M_{a2} = -k_{a2}\varphi_2$$
$$M_{b2} = +\overline{m}_{b1} + k_{b1}\varphi_2 + k_{bi}\varphi_3$$
$$M_{b3} = -\overline{m}_{b2} - k_{bi}\varphi_2 - k_{b2}\varphi_3$$

In this way, the analysis has been brought to the isostatic beam of Fig. 3.8 on which one proceeds to the calculation of the reactions and the drawing of the diagrams of internal forces as made for the Force Method. In particular, following the equilibria imposed by the solving equations, it will be:

$$M_{a2} = M_{b2}$$
$$M_{b3} = 0$$

To be noticed that the sign conventions change when passing from the stiffness k and fixed end moments $\overline{m}$ to the end moments of the members; the former ones are taken positive if clockwise so to be directly summed up at the joints when writing the equilibrium equations; the latter ones are taken with the conventions of the related internal forces (e.g. positive if the lower side of the beams is tensioned).

What presented above is the application procedure of the *Displacement Method*, of which Fig. 3.9 shows the flow chart, evidencing the "duality" with the Force Method.

Fig. 3.9 Flow chart of the
procedure

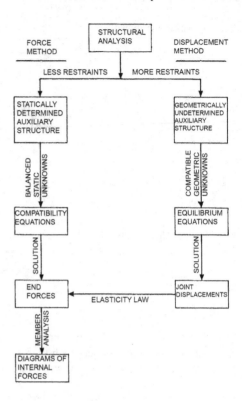

It can be noticed that the procedure has been prolonged by one step: in fact, assuming that the aim of the structural analysis is the definition of the internal forces, following the Force Method these come directly from the values of the hyperstatics (end forces of the members) as obtained through the solution of the solving equations; following the Displacement Method instead the solution of the solving equations provides the joint displacements, from which one has to pass to the end forces of members applying the pertinent elasticity law.

Despite the additional step, the Displacement Method is in general much more effective. First, it can be applied in an integral standardization of the procedures based on the analysis of only one type of member (the beam with two end full supports); then, one does not find the lability problems that arise with the Force Method when the redundant restraints are deleted in the joints of frame structures; again, there is not the burden of the global cinematic analysis for the evaluation of the effects of the axial deformation of the members since the additional restraints block at the joints any mutual action so that also the axial behaviour can be analysed separately for any single member.

For these reasons, the Displacement Method, following the standard procedure that will be presented at the subsequent section, is universally adopted in the auto-mated calculation codes of structures. For manual applications, to the Force Method

Fig. 3.10 Shortened
analysis of the continuous
beam

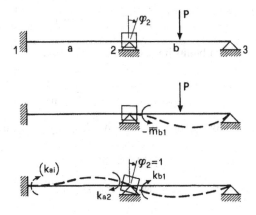

can be reserved some preference sector, such as that of the continuous beams, considering that both the methods gain their operative effectiveness from the preventive availability of the pertinent elastic coefficients: for the Force Method, the expressions of the flexibilities are needed as deduced at Sect. 2.1.2; for the Displacement Method, the expressions of the stiffness are needed as deduced at Sect. 2.2.1 of the preceding chapter.

This introduction concludes showing as, for a manual calculation, also with the Displacement Method one can save equations in the solving algorithm, simply widening the set of the tabled coefficients. Similarly to what done with the Force Method for the structure of Fig. 2.43 using the flexibility of the beam supported with a full restraint and a simple bearing, the stiffness of the same member can be used to omit, in the analysis of the continuous beam of concern, the second geometric unknown in the equilibrium equation.

Figure 3.10 shows such procedure that leads to only one equilibrium equation:

$$m_{22}\varphi_2 + m_{2o} = 0$$

with

$$m_{22} = k_{a2} + k_{b1}$$
$$m_{2o} = \overline{m}_{b1}$$

where in particular, with members with constant section, referring to the cases of Figs. 2.28, 2.34 and 2.85, one has:

$$k_{a2} = \frac{4EI_a}{l_a}$$
$$k_{b1} = \frac{3EI_b}{l_b}$$

$$\overline{m}_{b1} = -\frac{3Pl_b}{16}$$

One obtains immediately the solution:

$$\varphi_2 = -\frac{m_{2o}}{m_{22}} = \frac{-\overline{m}_{b1}}{k_{a2} + k_{b1}}$$

and subsequently, through the elasticity law of the members:

$$(M_{b3} = 0)$$

$$M_{b2} = \overline{m}_{b1} + k_{b1}\varphi_2 = \overline{m}_{b1} + \overline{m}_{b1}\frac{k_{b1}}{k_{a2} + k_{b1}}$$

$$M_{a2} = -k_{a2}\varphi_2 = +\overline{m}_{b1}\frac{k_{a2}}{k_{a2} + k_{b1}} \quad (= M_{b2})$$

$$M_{a1} = +k_{ai}\varphi_2 = -\overline{m}_{b1}\frac{k_{ai}}{k_{a2} + k_{b1}}$$

Setting again $l_a = l_b = l$ and $I_a = I_b = I$, one eventually has:

$$\varphi_2 = -\frac{3Pl^2}{112EI}$$

that leads obviously to the same result

$$M_{a2} = -\frac{3}{28}Pl \quad (= M_{b2})$$

obtained at Sect. 2.2.2 with the Force Method.

3.1.2 Standard Procedure

A first example of application of the Displacement Method is shown with reference to the simple frame of Fig. 3.11a, with only one internal joint. In the example of continuous beam discussed at the previous section, the analysis was addressed only to the flexural behaviour of the beam and by consequence the additional supports represented rotational restraints. In the case of plane frames, the additional supports of the auxiliary structure shall prevent all the three components that define the displacement of a joint in the plane: two translations and one rotation. This refers obviously to the internal joints and not to the external joints with full supports: for these latter, the displacement remains identically null.

So, Fig. 3.11b shows the auxiliary structure where at the joint 1 the "wedge" corresponding to a full support has been added. The three unknowns corresponding

Fig. 3.11 Plane case:
a given frame, **b** auxiliary
structure

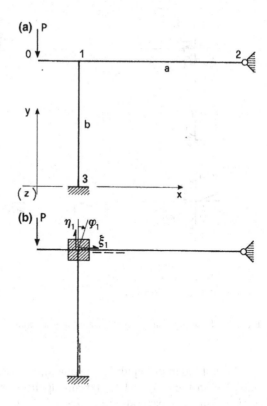

to the suppressed degrees of freedom have also been indicated. These unknowns are measured on the global reference system as:

– the translation ξ_1 in the direction of the axis X;
– the translation η_1 in the direction of the axis Y;
– the rotation φ_1 around the axis Z.

The solving algorithm is composed by one equilibrium equation to the horizontal translation, one equilibrium equation to the vertical translation and one equilibrium equation to the rotation:

$$\begin{cases} h_1 = 0 \\ v_1 = 0 \\ m_1 = 0 \end{cases}$$

where h_1 indicates the horizontal force, v_1 indicates the vertical force and m_1 indicates the moment. These three components refer to the reaction rendered by the additional support to the members that converge on it and are assumed positive if directed like the corresponding geometric unknowns.

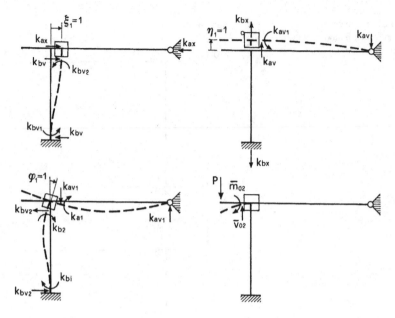

Fig. 3.12 Superposition of effects on auxiliary structure

The drafting of the equilibrium equations is based on the superposition of effects described in Fig. 3.12, where for any applied action the moments and the forces arising at the member ends are evidenced: the moments are oriented with the "tail" in the tensioned side of the deflected member, the shear forces are oriented so to give a couple of opposite direction with respect to the moments to which they provide equilibrium, and the axial forces are oriented consistently to the elongation–tension and shortening–compression of the members. So summing up the components at the concerned joint 1, one has:

$$\begin{cases} h_{1\xi}\xi_1 + h_{1\eta}\eta_1 + h_{1\varphi}\varphi_1 + h_{10} = 0 \\ v_{1\xi}\xi_1 + v_{1\eta}\eta_1 + v_{1\varphi}\varphi_1 + v_{10} = 0 \\ m_{1\xi}\xi_1 + m_{1\eta}\eta_1 + m_{1\varphi}\varphi_1 + m_{10} = 0 \end{cases}$$

with

$$\begin{aligned} &h_{1\xi} = +k_{ax} + k_{bv} \quad v_{1\xi} = 0 \qquad\qquad m_{1\xi} = -k_{bv2} \\ &h_{1\eta} = 0 \qquad\qquad\quad v_{1\eta} = +k_{av} + k_{bx} \quad m_{1\eta} = -k_{av1} \\ &h_{1\varphi} = -k_{bv2} \qquad\quad v_{1\varphi} = -k_{av1} \qquad\quad m_{1\varphi} = +k_{a1} + k_{b2} \\ &h_{10} = 0 \qquad\qquad\quad v_{10} = +\bar{v}_{02} \qquad\quad m_{10} = \bar{m}_{02} \end{aligned}$$

Fig. 3.13 Orientation of the member

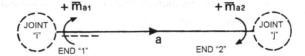

In particular, one can notice how the cantilever 0–1 constitutes an isostatic protrusion, with null stiffness, that transmits its invariant action to the joint: its static behaviour does not depend on the unknowns of the system.

To avoid confusions in the symbols used, it is specified that the subscripts 1 and 2 set to the stiffness and end forces refer to the "first" and "second" end of the members as oriented in the static scheme. The orientation can be set marking with a dashed line the side tensioned by the positive bending moments or with an arrow as indicated in Fig. 3.13. The subscripts set to the joint quantities (like ξ_j, η_j, ..., m_j) refer instead to the progressive numbering of the joints.

For members with constant cross section, the equations become:

$$\begin{cases} \left(\dfrac{EA_a}{l_a} + \dfrac{12EI_b}{l_b^3}\right)\xi_1 + (0\eta_1) - \dfrac{6EI_b}{l_b^2}\varphi_1 = 0 \\[2ex] (0\xi_1) + \left(\dfrac{3EI_a}{l_a^3} + \dfrac{EA_b}{l_b}\right)\eta_1 - \dfrac{3EI_a}{l_a^2}\varphi_1 = -P \\[2ex] -\dfrac{6EI_b}{l_b^2}\xi_1 - \dfrac{3EI_a}{l_a^2} + \left(\dfrac{3EI_a}{l_a} + \dfrac{4EI_b}{l_b}\right)\varphi_1 = -Pl_0 \end{cases}$$

where the coefficients have been deduced from the expressions of the stiffness recalled at the beginning of the present section. This equilibrium equation system is obviously equipollent to the compatibility one set through a more laborious way at Sect. 2.3.1.

The example elaborated here above belongs to the standard procedure of the Displacement Method. It is based on the addition of full supports at the joints and on the application of the geometric unknowns as many, for each joint, as were its degrees of freedom in the given structure (e.g. three for the plane frames, six for the spatial frames). Also, the axial deformations are considered through which the axial forces are analysed in each member separately.

This procedure allows a complete standardization of the calculation processes, with a high number of unknowns. If one adopts the simplifications allowed by the negligible influence of the axial deformations, with the approximations stated at the quoted Sect. 2.3.1, the solving algorithms can be much reduced in equation number. If, for example, the axial stiffness of the members is set infinite (with $k_{ax} = k_{bx} = \infty$), one obtains null translations:

$$\xi_1 = 0$$
$$\eta_1 = 0$$

The only third equation remains that expresses the rotation equilibrium of the joint (with $\overline{m}_{02} = Pl_0$):

$$(k_{a1} + k_{b2})\varphi_1 = -\overline{m}_{02}$$

and leads to

$$\varphi_1 = \frac{-\overline{m}_{02}}{k_{a1} + k_{b2}}$$

and to the end moments:

$$M_{a1} = +k_{a1}\varphi_1 = -\overline{m}_{02}\frac{k_{a1}}{k_{a1} + k_{b2}}$$

$$M_{b1} = -k_{b2}\varphi_1 = -\overline{m}_{02}\frac{k_{b2}}{k_{a1} + k_{b2}}$$

$$M_{b3} = k_{bi}\varphi_1 = -\overline{m}_{02}\frac{k_{bi}}{k_{a1} + k_{b2}}$$

In the case of members with constant cross section and with $l_a = l_b = l$ and $I_a = I_b = I$, one obtains again the solution:

$$M_{a1} = -\frac{3}{7}Pl_0$$

$$M_{b1} = +\frac{4}{7}Pl_0$$

$$M_{b3} = -\frac{2}{7}Pl_0$$

from which one has the diagrams already shown in Fig. 2.60 of Sect. 2.3.1.

3.1.3 Transformation of Co-ordinate Axes

The preceding example referred to a frame with orthogonal members for which the vectorial sum of the joint forces did not require previous projections of the components. Assuming, with reference to the equilibrium equations, a global system with the axes X and Y parallel to the sides of the structural mesh and the axis Z orthogonal to it, one had to sum up algebraically the axial forces of the horizontal members and the shear forces of the vertical members for the first translation equilibrium, and one had to sum up algebraically the shear forces of the horizontal members and the axial forces of the vertical members for the second translation equilibrium.

Also, the geometric unknowns were parallel to the sides of the structural mesh, so that for a given translation one had pure axial deformations for the members oriented

Fig. 3.14 Projection of
forces and displacements

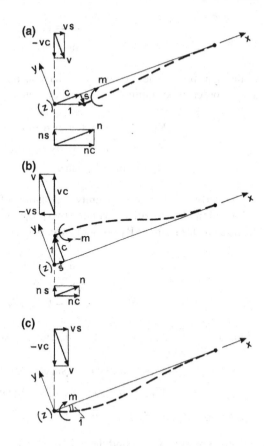

in its direction and pure flexural deformations for the orthogonal members. The
elasticity law of the members, based on the stiffness that, as shown in the preceding
sections, is measured on the local system of axes, gave directly the components of
the joint forces in the directions of the equilibrium equations.

In the general case of frames with inclined members, such as that shown in Fig. 3.5,
the formulation of the solving algorithm requires instead proper projections of the
forces and displacements from one system of axes to the other.

Let's consider, for example, the response rendered by the member with an incli-
nation α over the horizontal axis. For the application of the components ξ_i, η_i, φ_i
of the joint displacement, one has on the same joint the forces described in detail in
Fig. 3.14, where in particular it is set $c = \cos\alpha$ and $s = \sin\alpha$.

For the horizontal unity translation, the member of concern deforms as indicated
in Fig. 3.14a, displaying a shortening equal to $\cos\alpha$ and an orthogonal deflection
equal to $\sin\alpha$. These deformation components, measured on the local system of axes,
lead to the components

$$n = k_x \cos \alpha$$

$$v = k_v \sin \alpha$$
$$m = k_{v1} \sin \alpha$$

of the joint force. These components have to be projected on the global system of axes in order to be summed up in the pertinent equilibrium equations:

$$K_{XX} = n \cos \alpha + v \sin \alpha = k_x \cos^2 \alpha + k_v \sin^2 \alpha$$
$$K_{YX} = n \sin \alpha - v \cos \alpha = (k_x - k_v) \cos \alpha \sin \alpha$$
$$K_{ZX} = m = k_{v1} \sin \alpha$$

Similarly, for the vertical unity translation of the joint, Fig. 3.14b shows the deformation of the member, with the two projections axial $\sin\alpha$ and orthogonal $\cos\alpha$ of the same translation. By consequence, one has the components

$$n = k_x \cos \alpha$$
$$v = k_v \cos \alpha$$
$$m = k_{v1} \cos \alpha$$

of the joint force that, projected on the global system of axes, give:

$$K_{XY} = n \cos \alpha - v \sin \alpha = (k_x - k_v) \cos \alpha \sin \alpha$$
$$K_{YY} = n \sin \alpha + v \cos \alpha = k_x \sin^2 \alpha + k_v \cos^2 \alpha$$
$$K_{ZY} = m = k_{v1} \cos \alpha$$

The unity rotation around the axis Z, as indicated in Fig. 3.14c, does not require any transformation since the two axes z and Z coincide. So, one has the components

$$n = 0$$
$$v = k_{v1}$$
$$m = k_1$$

of the joint force that, transferred to the global system of axes, lead to

$$K_{XZ} = v \sin \alpha = k_{v1} \sin \alpha$$
$$K_{YZ} = -v \cos \alpha = -k_{v1} \cos \alpha$$
$$K_{ZZ} = m = k_1$$

In order to perform this *transformation of co-ordinates* with a double-*axis rotation*, repeated also for the second end of the member and for the indirect reciprocal effects, the operations can be expressed with the algorithms of the matrix algebra. So, with reference again to the first end of the member, a matrix k is formulated, containing in a regular format the original stiffness terms:

$$
\begin{array}{c@{\quad}ccc}
 & \delta_x & \delta_y & \varphi_z \\
n & k_x & 0 & 0 \\
v & 0 & k_v & -k_{v1} \\
m & 0 & -k_{v1} & k_1
\end{array}
$$

where the rows are referred to the single components axial, orthogonal (shear) and rotational (moment) of the joint force, the columns to the corresponding displacement components.

The transfer of this matrix to the global system of axes, to obtain the new matrix K:

$$
\begin{array}{c@{\quad}ccc}
 & \xi & \eta & \zeta \\
r_X & K_{XX} & K_{XY} & K_{XZ} \\
r_Y & K_{YX} & K_{YY} & K_{YZ} \\
m_Z & K_{ZX} & K_{ZY} & K_{ZZ}
\end{array}
$$

is obtained through a double product with the "transfer matrix" T that contains the direction cosines of the axes of the local system with respect to those of the global system:

$$
\begin{array}{c@{\quad}ccc}
 & x & y & z \\
X & \cos\alpha & -\sin\alpha & 0 \\
Y & \sin\alpha & \cos\alpha & 0 \\
Z & 0 & 0 & 1
\end{array}
$$

From these operations, one obtains the expressions presented before.

The procedures of the automated computation are based on such approach. For more details, reference shall be made to the pertinent disciplines. The present chapter presents only the problem through few notes that indicate its application. Anyway, one can notice the laboriousness of the procedures if used within a hand calculation. At Sect. 3.3.3, one can better evaluate the complexity of the numerical calculations required for the analysis of frames with inclined members, also when the simplified procedures are used with reduced number of equilibrium equations. In the case of the frame of Fig. 3.5, without considering the computation burdens of the co-ordinate transformation, the standard procedure would require the solution of a system of six equilibrium equations.

Another problem of co-ordinate transformation refers to the finite dimensions of joints, with converging members variously located with respect to the "centre" chosen for the calculation of the moments in the corresponding equilibrium equation. Figure 3.15 shows a possible structural situation that presents this problem.

Fig. 3.15 Example of joint
finite dimensions

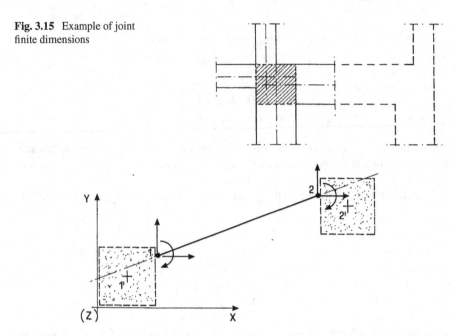

Fig. 3.16 Example of beam connected to rigid bodies

The Displacement Method for these cases employs a frame scheme constituted by an assembly of ordinary beams connected to non-deformable bodies of finite dimensions. The current flexural and axial deformability of the beams is extended up to the end sections 1 and 2 of the beams (Fig. 3.16); so, their behaviour is represented by the stiffness of these sections that, in the quoted figure, are drawn already rotated to the global system of axes following the transformation of co-ordinates presented before.

Considering the joint connected with the first end of the beam, in order to sum up the components in their respective equations, one has to transfer the end quantities of the beam from its end 1 to the point 1′ chosen as "centre" of the joint.

In Fig. 3.17a, the geometric quantities are represented, that is translations and rotation; to deduce the parameters of the elastic behaviour of the beam, referred to its end 1, from the unknowns of the equilibrium equations, referred to the centre 1′, one has the following relations:

$$\xi = \xi' + e_y \varphi'$$
$$\eta = \eta' - e_x \varphi'$$
$$\varphi = \varphi'$$

Fig. 3.17 Geometric quantities

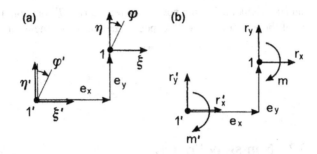

The elasticity law gives the corresponding components of the end forces of the beam that, with the symbols introduced before, are:

$$r_X = K_{XX}\,\xi + K_{XY}\,\eta + K_{XZ}\,\varphi$$
$$r_Y = K_{YX}\,\xi + K_{YY}\,\eta + K_{YZ}\,\varphi$$
$$m = K_{ZX}\,\xi + K_{ZY}\,\eta + K_{ZZ}\,\varphi$$

These components have to be transferred back to the centre $1'$ of the joint (Fig. 3.17b) with:

$$r'_X = r_X$$
$$r'_Y = r_Y$$
$$m' = m + e_Y r_X - e_X r_Y$$

With a double substitution, the elasticity law is in this way transferred to the centre of the joint, obtaining the stiffness terms

$$K'_{XX} = K_{XX}$$
$$K'_{YX} = K_{YX}$$
$$K'_{ZX} = K_{ZX} + e_Y K_{YY} - e_X K_{YX}$$
$$K'_{XY} = K_{XY}$$
$$K'_{YY} = K_{YY}$$
$$K'_{ZY} = K_{ZY} + e_Y K_{XY} - e_X K_{YY}$$
$$K'_{XZ} = K_{XZ} + e_Y K_{XX} - e_X K_{XY}$$
$$K'_{YZ} = K_{YZ} + e_Y K_{YX} - e_X K_{YY}$$
$$K'_{ZZ} = K_{ZX} + e_Y (K_{ZX} + e_Y K'_{XZ}) - e_X (K_{ZY} + e_X K'_{YZ})$$

Also, this *co-ordinate transformation* with a double-*axis translation*, repeated for the second end of the beam and for the indirect effects, can be expressed by a double-matrix product using the new "transfer matrix" T' that contains the eccentricities e_X, e_Y of the end section with respect to the centre of the joint:

$$\begin{vmatrix} 1 & 0 & 0 \\ 0 & 1 & 0 \\ e_Y & -e_X & 1 \end{vmatrix}$$

3.2 Non-sway Frames

For the non-sway frames as defined in Sect. 2.3.2, the neglect of the axial deformations of the members leads to null translation components of the joint displacements. Within the applicability field of such approximation, as defined in Sect. 2.3.1, the Displacement Method is applied with the only rotation unknowns, one for each joint in the case of plane frames and three for each joint in the case of spatial frames.

In order to show the first application of the method, let's consider the same frame of Fig. 2.61, already analysed with the Force Method at Sect. 2.3.2 with the three static unknowns evidenced in Fig. 2.61c. With the Displacement Method, the geometric unknowns remain two, corresponding to the number of the internal joints for any possible number of converging members.

The concerned frame is shown in Fig. 3.18a, with a new load and having specified the measure of the involved quantities. Figure 3.18b shows the auxiliary structure, with the rotation restraints added to the joint 1 and 2 and marking the corresponding unknowns φ_1 and φ_2.

The set of equilibrium equations

$$\begin{cases} m_{11}\varphi_1 + m_{12}\varphi_2 + m_{10} = 0 \\ m_{21}\varphi_1 + m_{22}\varphi_2 + m_{20} = 0 \end{cases}$$

is written with the superposition of effects shown in Fig. 3.19. With the recurrent expressions of the stiffness of the beams with constant section, these equations are:

$$\begin{cases} \left(\dfrac{4EI}{1} + \dfrac{4EI'}{1}\right)\varphi_1 + \dfrac{2EI'}{1}\varphi_2 - \dfrac{pl^2}{12} = 0 \\ \dfrac{2EI'}{1}\varphi_1 + \left(\dfrac{4EI'}{1} + \dfrac{3EI'}{1} + \dfrac{4EI}{1}\right)\varphi_2 + \dfrac{pl^2}{12} = 0 \end{cases}$$

and, after the pertinent simplifications (with $I' = 2I$), become

$$\begin{cases} 6\varphi_1 + 2\varphi_2 = \dfrac{pl^3}{24EI} \\ 2\varphi_1 + 9\varphi_2 = \dfrac{-pl^3}{24EI} \end{cases}$$

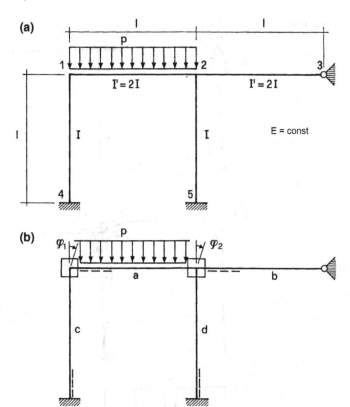

Fig. 3.18 Example of plane frame

The solution is

$$\varphi_1 = +\frac{11}{50}\frac{pl^3}{24EI}$$

$$\varphi_2 = -\frac{8}{50}\frac{pl^3}{24EI}$$

and from this solution, through the elasticity relations, the end moments of the members are obtained:

$$M_{a2} = -\frac{2EI'}{l}\varphi_1 - \frac{4EI'}{l}\varphi_2 - \frac{pl^2}{12} = -\frac{40}{50}\frac{pl^2}{12}$$

$$M_{b2} = +\frac{3EI'}{l}\varphi_2 = -\frac{24}{50}\frac{pl^2}{12}$$

$$M_{c1} = -\frac{4EI}{l}\varphi_1 = -\frac{22}{50}\frac{pl^2}{12}$$

Fig. 3.19 Superposition of effects

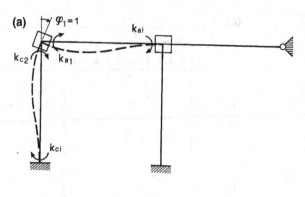

(a)

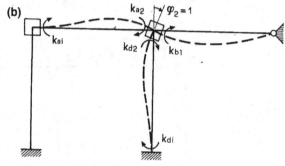

(b)

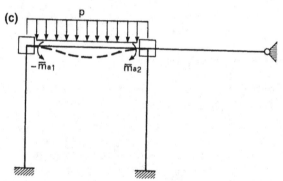

(c)

$$M_{c4} = -M_{c1}/2 = +\frac{11}{50}\frac{pl^2}{12}$$

$$M_{d2} = -\frac{4EI}{1}\varphi_2 = +\frac{16}{50}\frac{pl^2}{12}$$

$$M_{d5} = -M_{d2}/2 = -\frac{8}{50}\frac{pl^2}{12}$$

In particular, the equilibria at joints 1 and 2, as set by the same solving equations, are satisfied being:

$$-M_{c1} + M_{a1} = 0$$
$$-M_{a2} + M_{b2} - M_{d2} = 0$$

Eventually, the problem has been led to the isostatic structure of Fig. 3.20 on which one can follow the same procedure of the Force Method. Figure 3.21 gives the diagrams of the internal forces for which in particular one has:

$$M_a = +\frac{pl^2}{8} + \frac{M_{a1} + M_{a2}}{2} = +\frac{44}{50}\frac{pl^2}{12}$$

$$V_{a1} = +\frac{pl}{2} + \frac{M_{a2} - M_{a1}}{1} = +\frac{282}{300}\frac{pl}{2}$$

$$V_{a2} = -\frac{pl}{2} + \frac{M_{a2} - M_{a1}}{1} = -\frac{318}{300}\frac{pl}{2}$$

$$V_b = -\frac{M_{b2}}{1} = +\frac{24}{300}\frac{pl}{2}$$

$$V_c = +\frac{M_{c1} - M_{c4}}{1} = -\frac{33}{300}\frac{pl}{2}$$

$$V_d = +\frac{M_{d2} - M_{d5}}{1} = +\frac{24}{300}\frac{pl}{2}$$

For the calculation of axial forces, having neglected the axial deformations of the members, a separate analysis member by member is not possible; the global analysis of the isostatic structure of Fig. 3.20 is necessary. Due to the orthogonal disposition of the members, this analysis is easily performed with:

$$N_a = V_c = -\frac{33}{300}\frac{pl}{2}$$

$$N_b = N_a + V_d = -\frac{9}{300}\frac{pl}{2}$$

$$N_c = -V_{a1} = -\frac{282}{300}\frac{pl}{2}$$

$$N_d = +V_{a2} - V_b = -\frac{342}{300}\frac{pl}{2}$$

Fig. 3.20 Equivalent isostatic structure

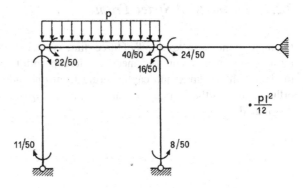

Fig. 3.21 Diagrams of
internal forces

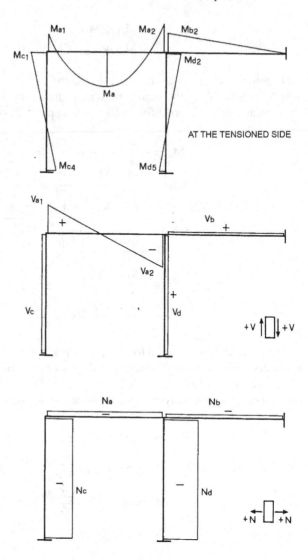

3.2.1   *Example of Space Frame*

Figure 3.22 shows an "excerpt" of three-dimensional frame structure with orthogonal
mesh: one vertical column connected at its top with two horizontal beams. One of
the beams holds up a slab cantilevering from one side, and the other is oriented
orthogonally to the former. On the slab, a load q of constant intensity (N/m^2) is
distributed.

Fig. 3.22 Excerpt of space
frame

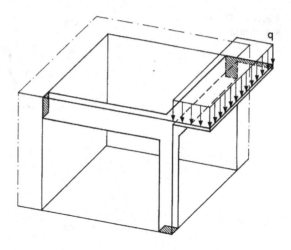

Figure 3.23a shows the static scheme of this structure, with three members orthogonal to each other. The beams are laid so to have a vertical plane of uniaxial bending, as indicated by the local axes y, z of their cross sections, and the column is laid so to have the inertia principal axes of its cross section contained in the quoted vertical planes. This member arrangement allows to formulate the equilibrium equations without projections of forces and displacements.

In Fig. 2.23b, the global system of axes X, Y, Z is indicated, chosen consistently to the structural layout, with the sign conventions for the joint quantities (rotations and moments). Full supports are placed at the external joints 2, 3 and 4. In the same figure is also indicated the load, brought to the central longitudinal axis of the involved beam, with its components flexural p and torsional f:

$$p = q\,b \quad (\text{N/m})$$
$$f = p\,e \quad (\text{N m/m})$$

with b projection of the slab and e eccentricity of the load from the beam axis.

The stress analysis of the frame is performed neglecting the axial deformations of the members. This leads to consider as geometric unknowns the only three rotation components of the displacement of the joint 1, setting the correspondent equilibrium equations:

$$\begin{cases} m_{XX}\varphi_X + m_{XY}\varphi_Y + m_{XZ}\varphi_Z + m_{X0} = 0 \\ m_{YX}\varphi_X + m_{YY}\varphi_Y + m_{YZ}\varphi_Z + m_{Z0} = 0 \\ m_{ZX}\varphi_X + m_{ZY}\varphi_Y + m_{ZZ}\varphi_Z + m_{Z0} = 0 \end{cases}$$

Figure 3.24 shows the superposition of effects. One can notice that the orthogonal structural arrangement leads to three independent equations, due to the null value of all the indirect coefficients. Using the indexes a, b, c with reference to the members,

Fig. 3.23 Details of the
static scheme

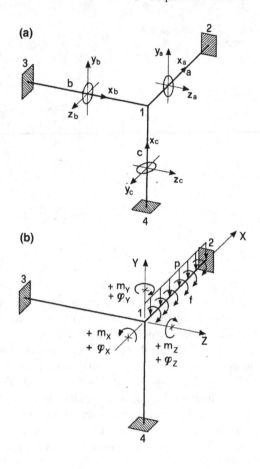

(a)

(b)

the symbol k_t for the torsional stiffness, the symbols k_y, k_z for the direct rotation
stiffness (of the pertinent end), respectively, for the rotations around y and around z,
and the symbols $\bar{t}'_a$, $\bar{t}''_a$, $\bar{m}'_{az}$, $\bar{m}''_{az}$ for the full support moments at the two ends of the
loaded beam, one has:

$$\begin{cases} (k_{at} + k_{bz} + k_{cy})\varphi_X + \bar{t}'_a = 0 \\ (k_{ay} + k_{by} + k_{ct})\varphi_Y = 0 \\ (k_{az} + k_{bt} + k_{cz})\varphi_Z + \bar{m}'_{az} = 0 \end{cases}$$

The independence of the three equilibrium equations corresponds to the behaviour
of orthogonal frames one independent from the others. Therefore, neglecting the
horizontal frame which is not loaded, one can analyse the vertical plane frames 3–1–4
and 4–1–2 in a separate way, cumulating the effects on the column that belongs to
both the quoted frames. The two orthogonal analyses, in the simple case of concern,
are performed, respectively, with:

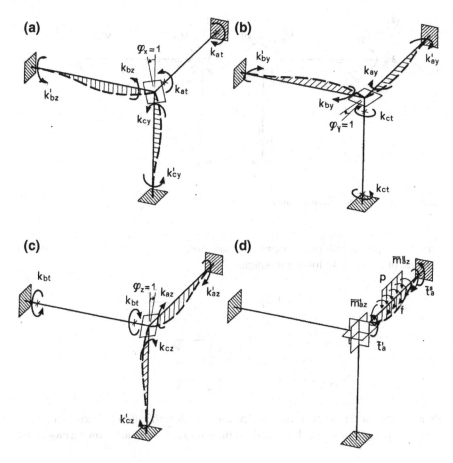

Fig. 3.24 Superposition of effects

$$\varphi_X = \frac{-\bar{t}'_a}{k_{at} + k_{bz} + k_{cy}}$$

$$\varphi_Z = \frac{-\bar{m}'_{az}}{k_{az} + k_{bt} + k_{cz}}$$

For the first frame (see Fig. 3.25a) at the joint 1, one has the following moments:

$$T_{a1} = -\bar{t}'_a - k_{at}\varphi_x = -\bar{t}'_a + \bar{t}'_a \frac{k_{at}}{k_{at} + k_{bz} + k_{cy}}$$

$$M_{b1z} = k_{bz}\varphi_x = -\bar{t}'_a \frac{k_{bz}}{k_{at} + k_{bz} + k_{cy}}$$

$$M_{c1y} = k_{cy}\varphi_x = -\bar{t}'_a \frac{k_{cy}}{k_{at} + k_{bz} + k_{cy}}$$

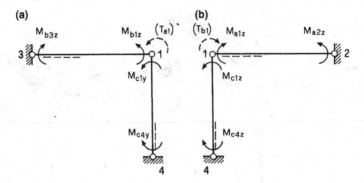

Fig. 3.25 Two orthogonal plane frames

and at the opposite ends of the members, indicating with k'_y, k'_z the pertinent indirect stiffness, one has the following moments:

$$T_{a2} = +\overline{t}''_a - k_{at}\varphi_x = +\overline{t}''_a + \overline{t}'_a \frac{k_{at}}{k_{at} + k_{bz} + k_{cy}}$$

$$M_{b3z} = -k'_{bz}\varphi_x = \overline{t}'_a \frac{k'_{bz}}{k_{at} + k_{bz} + k_{cy}}$$

$$M_{c4y} = -k'_{cy}\varphi_x = \overline{t}'_a \frac{k'_{cy}}{k_{at} + k_{bz} + k_{cy}}$$

where the torsional moment is assumed positive if oriented in the clockwise screwing.
For the second frame (see Fig. 3.25b) at the joint 1, one has the following moments:

$$M_{a1z} = +\overline{m}'_{az} - k_{az}\varphi_z = +\overline{m}'_{az} - \overline{m}'_{az} \frac{k_{az}}{k_{az} + k_{bt} + k_{cz}}$$

$$T_{b1} = +k_{bt}\varphi_z = -\overline{m}'_{az} \frac{k_{bt}}{k_{az} + k_{bt} + k_{cz}}$$

$$M_{c1z} = -k_{cz}\varphi_z = +\overline{m}'_{az} \frac{k_{cz}}{k_{az} + k_{bt} + k_{cz}}$$

and at the opposite ends of the members, one has the following moments:

$$M_{a2z} = -\overline{m}''_{az} - k'_{az}\varphi_z = -\overline{m}''_{az} + \overline{m}'_{az} \frac{k'_{az}}{k_{az} + k_{bt} + k_{cz}}$$

$$T_{b3} = k_{bt}\varphi_z = -\overline{m}'_{az} \frac{k'_{bt}}{k_{az} + k_{bt} + k_{cz}}$$

$$M_{c4z} = k'_{cz}\varphi_z = -\overline{m}'_{az} \frac{k'_{cz}}{k_{az} + k_{bt} + k_{cz}}$$

For each plane analysis, farther to the bending actions contained in the plane of the frame, also the torsional actions coming from the orthogonal beam have been introduced.

The analysis is now completed in the case of members with constant cross section and with

$$l_a = l_b = l_c = 1$$
$$I_{cy} = I_{cz} = I$$
$$I_{az} = I_{bz} = 4\,I$$
$$G/E = 2/5$$
$$J_a = J_b = 5\,I$$

The fixed end moments of the beam a are equal to:

$$\bar{m}'_{az} = -\frac{pl^2}{12}$$
$$\bar{m}''_{az} = +\frac{pl^2}{12}$$
$$\bar{t}'_a = +\frac{fl}{2}$$
$$\bar{t}''_a = +\frac{fl}{2}$$

The stiffness are (with $k_o = EI/l$):

$$k_{az} = \frac{4EI_{az}}{l_a} = 16k_o \quad (k'_{az} = 8k_o)$$
$$k_{at} = \frac{GJ_a}{l_a} = 2k_o$$
$$k_{bz} = \frac{4EI_{bz}}{l_b} = 16k_o \quad (k'_{bz} = 8k_o)$$
$$k_{bt} = \frac{GJ_b}{l_b} = 2k_o$$
$$k_{cy} = \frac{4EI_{cy}}{l_c} = 4k_o \quad (k'_{cy} = 2k_o)$$
$$k_{cz} = \frac{4EI_{cz}}{l_c} = 4k_o \quad (k'_{cz} = 2k_o)$$

and lead, with $\overline{T} = fl/2$ and $\overline{M} = pl^2/12$, to the solution:

$$\varphi_x = -\frac{1}{22}\frac{\overline{T}}{k_o}$$

$$\varphi_z = +\frac{1}{22}\frac{\overline{M}}{k_o}$$

In the plane of the frame 3–1–4, the end moments of the members indicated in Fig. 3.25a are:

$$T_{a1} = -\overline{T}\left(1 - \frac{2}{22}\right) = -\frac{10}{11}\overline{T}$$

$$M_{b1z} = -\overline{T}\frac{16}{22} = -\frac{8}{11}\overline{T}$$

$$M_{c1y} = -\overline{T}\frac{4}{22} = -\frac{2}{11}\overline{T}$$

$$T_{a2} = +\overline{T}\left(1 + \frac{2}{22}\right) = +\frac{12}{11}\overline{T}$$

$$M_{b3z} = +\overline{T}\frac{8}{22} = +\frac{4}{11}\overline{T}$$

$$M_{c4y} = +\overline{T}\frac{2}{22} = +\frac{1}{11}\overline{T}$$

In Fig. 3.26, the corresponding diagrams of the internal forces are shown, for which in particular one has:

$$V_{by} = \frac{M_{b1z} - M_{b3z}}{1} = -\frac{12}{11}\frac{\overline{T}}{1} = N'_c$$

$$V_{cz} = \frac{M_{c1y} - M_{c4y}}{1} = -\frac{3}{11}\frac{\overline{T}}{1} = -N_b$$

In the plane of the frame 4–1–2, the end moments of the members indicated in Fig. 3.25b are:

$$M_{a1z} = -\overline{M}\left(1 - \frac{16}{22}\right) = -\frac{3}{11}\overline{M}$$

$$T_{b1} = +\overline{M}\frac{2}{22} = +\frac{1}{11}\overline{T} = T_{b3}$$

$$M_{c1z} = -\overline{M}\frac{4}{22} = -\frac{2}{11}\overline{M}$$

$$M_{a2z} = -\overline{M}\left(1 + \frac{8}{22}\right) = -\frac{15}{11}\overline{M}$$

$$M_{az} = +\frac{3}{2}\overline{M} + \frac{M_{a1z} + M_{a2z}}{2} = +\frac{15}{22}\overline{M}$$

$$M_{c4z} = +\overline{M}\frac{2}{22} = +\frac{1}{11}\overline{M}$$

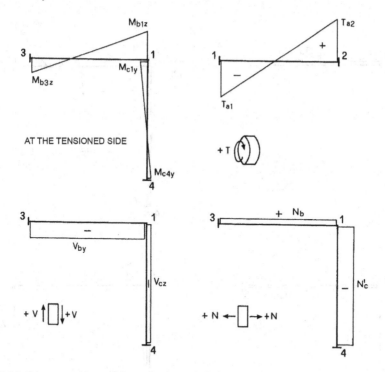

Fig. 3.26 Diagrams of internal forces for frame 3–1–4

In Fig. 3.27, the corresponding diagrams of the internal forces are shown, for which in particular one has (with $\overline{V} = pl/2$):

$$V_{a1y} = +\overline{V} + \frac{M_{a2z} - M_{a1z}}{l} = -\frac{9}{11}\overline{V} = -N_c''$$

$$V_{a2y} = -\overline{V} + \frac{M_{a2z} - M_{a1z}}{l} = +\frac{13}{11}\overline{V}$$

$$V_{cy} = \frac{M_{c1z} - M_{c4z}}{l} = -\frac{1}{22}\overline{V} = N_a$$

$$(N_c = N_c' + N_c'')$$

3.2.2 Moment Repartition Factors

With reference to the examples shown up to here, an operative interpretation of the Displacement Method can be deduced. This interpretation gives firstly the direct evaluation of the repartition of a given moment among the members convergent to

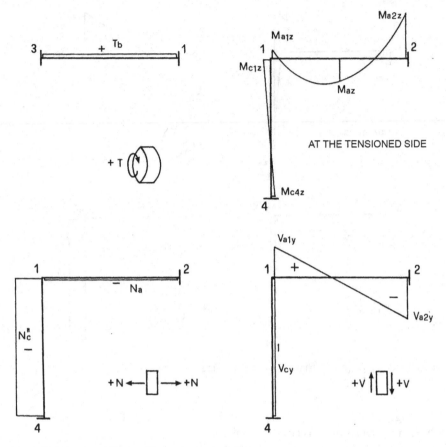

Fig. 3.27 Diagrams of internal forces for frame 4–1–2

the joint on which it is applied. A particular numeric procedure is also obtained for the solution of the non-sway frames.

Let's consider the solution of the continuous beam of Fig. 3.10 as elaborated at the end of Sect. 3.1.1:

$$M_{a2} = +\overline{m}_{b1} \frac{k_{a2}}{k_{a2} + k_{b1}}$$

$$M_{b2} = -\overline{m}_{b1} \frac{k_{b1}}{k_{a2} + k_{b1}} + \overline{m}_{b1}$$

These expressions show that the non-balanced moment m_{20} ($=\overline{m}_{b1}$), calculated at the additional restraint of joint 2, is shared between the two members in parts proportional to the respective stiffness. The moments deduced from this repartition are to be summed up, for any member, to the previous full support moments possibly

present. Eventually at the opposite ends of the members, pertinent moments are
transmitted with:

$$M_{a1} = -\overline{m}_{b1}\frac{k_{ai}}{k_{a2}+k_{b1}} = M_{a2}\frac{k_{ai}}{k_{a2}}$$

$$(M_{b3} = 0)$$

these latter being evaluated, starting from the repartition ones through the transmission factor

$$t_{12} = \frac{k_{ai}}{k_{a2}}$$

This factor, reminding the expressions of the stiffness listed at the beginning of
Sect. 3.1, coincides with that d_{ai}/d_{a1} already defined at Sect. 2.2.1 (see Fig. 2.26).

With this introduction, the solution of the continuous beam of Fig. 3.10 can be
obtained with a numerical procedure operating directly on the sketch of the static
scheme, as described in Fig. 3.28. In the procedure, one should refer only to the
convention of the joint quantities that assume positive the clockwise moments. These
positive moments tension the lower side of the beams if placed at their left end, and
they tension the upper side if placed at their right end.

So, the operations develop following the sequence shown here after. The two
spans of the beam are assumed of constant cross section and of the same length.

– With reference to joint 2 (see Fig. 3.28) of which the equilibrium is to be set, as a
  first step the stiffness k_j of the converging members is evaluated and, dividing them
  by their sum K_2, the corresponding *repartition factors* $r_{j2} = k_j/K_2$ are calculated:

member	k_j	r_{j2}
"a"	$4EI/l : 7EI/l$	$= 0.57$
"b"	$3EI/l : 7EI/l$	$= 0.43$
total $K_2 = 7EI/l$		1.00

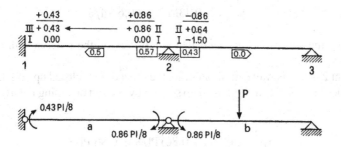

Fig. 3.28 Numerical procedure of moment repartition

The repartition factors, calculated in this way, are marked on the static scheme of the beam close to the pertinent ends.

- Subsequently, the *transmission factors* are evaluated on the basis of the supports present at the opposite ends of the beams, marking them on the static scheme together with the running direction:

$$t_{12} = 0.5$$
$$t_{32} = 0.0$$

- Having marked these factors, the procedure is started with the evaluation of the *fixed end moments* induced by loads at the ends of the members:

$$\overline{m}_{a1} = \overline{m}_{a2} = 0$$
$$\overline{m}_{b2} = -\frac{3Pl}{16} = -1.50Pl/8$$
$$(\overline{m}_{b3} = 0)$$

These values are written on the pertinent ends with code "I".

- All the fixed end moments are then summed up in the considered joint obtaining the *non-balanced residual* $M_2 = \Sigma \overline{m}_{j2}$:

$$M_2 = \overline{m}_{a2} + \overline{m}_{b2} = -1.50 \, Pl/8$$

- In order to restore the equilibrium, one changes the sign of the residual obtaining the *balancing moment*:

$$-M_2 = +1.5 \, Pl/8$$

- This moment is shared among the converging members by means of the respective factors r_{j2}:

$$m'_{a2} = -0.57M_2 = +0.86Pl/8$$
$$m'_{b2} = -0.43M_2 = +0.64Pl/8$$

obtaining the *repartition moments* that are written on the scheme with the code "II".

- The joint 2 is in this way equilibrated; so, the procedure is closed up with lines and the values "I" and "II" are summed up section by section obtaining the *hyperstatic moments*:

$$m_{a2} = (+0.00 + 0.86) \, Pl/8 = 0.86 \, Pl/8$$
$$m_{b2} = (-1.50 + 0.64) \, Pl/8 = -0.86 \, Pl/8$$

– Now, the values "II" are transmitted to the opposite ends of the members by means of the pertinent factors:

$$m''_{a1} = 0.5m_{a2} = +0.43Pl/8$$
$$(m''_{b3} = 0)$$

The *transmission moments* calculated in this way are written on the scheme with the code "III".

– Also in this latter section, the values "I" and "III" are summed up obtaining eventually the moments at the external supports:

$$m_{a1} = (+0.00 + 0.43)\, Pl/8 = +0.43\, Pl/8$$

Another applicative example of the numerical procedure of concern is presented with reference to the frame of Fig. 3.11 (Sect. 3.1.2). The approximate solution that neglects the axial deformations of the members is assumed. Omitting the interlocutory statements, the scheme of Fig. 3.29 is set up, for which, with $l_a = l_b = l$ and $I_a = I_b = I$, one has:

$$
\begin{array}{llll}
\text{"a"} & 3EI/l : 7EI/l & = 0.43 \\
\text{"b"} & 4EI/l : 7EI/l & = \underline{0.57} \\
& \overline{7EI/l} & 1.00
\end{array}
$$

$$\overline{m}_{01} = +1.00Pl_o = M_1$$

For the rest, the operations are directly written on the sketch and lead to:

$$m_{a1} = -0.43\, P\, l_o$$
$$m_{b1} = -0.57\, P\, l_o$$
$$m_{b3} = -0.285\, P\, l_o$$

Also, the two orthogonal frames of the spatial frame of Fig. 3.23, the one analysed at the preceding Sect. 3.2.1, can be solved with simple moment repartitions operated directly on the static schemes. In this case, also the torsional stiffness of the orthogonal beam is added to the bending stiffness of the members of any single plane frame to compose the *stiffness of joint* 1. The orthogonal beam, in the schemes of Figs. 3.30 and 3.31, is represented by the rotation spring, on which the value -1.0 is marked for the transmission factor of the torsional moment.

Following the conventions indicated in Fig. 3.23b for the joint quantities, the counterclockwise moments are assumed positive in the plane of the frame 3–1–4 of Fig. 3.30, and the clockwise moments are assumed positive in the plane of the frame 4–1–2 of Fig. 3.31. For the two frames, with the same data used at Sect. 3.2.1, one has the repartition factors and the fixed end moment calculated here below.

Fig. 3.29 Moment
repartition for a simple frame

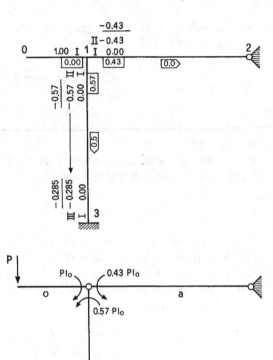

– Frame 3–1–4

$$\text{"a"} \quad GJ_a/l_a = 2EI/l : 22EI/l = 0.09$$
$$\text{"b"} \quad 4EI_b/l_b = 16EI/l : 22EI/l = 0.73$$
$$\text{"c"} \quad 4EI_c/l_c = \underline{4EI/l} : 22EI/l = \underline{0.18}$$
$$\qquad\qquad\qquad \overline{22EI/l} \qquad\qquad 1.00$$

$t_{a1} = +fl/2$ (counter clockwise torsional moment)

$t_{a2} = +fl/2$ (counter clockwise torsional moment)

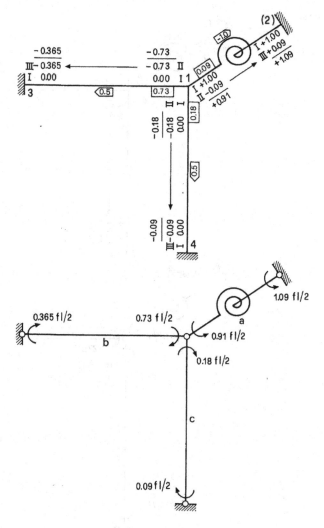

Fig. 3.30 Static scheme of frame 3–1–4

– Frame 4–1–2

$$"a" \ 4EI_a/l_a = 16EI/l : 22EI/l = 0.73$$
$$"b" \ GJ_b/l_b = 2EI/l : 22EI/l = 0.09$$
$$"c" \ 4EI_c/l_c = \underline{4EI/l} : 22EI/l = \underline{0.18}$$
$$\qquad\qquad\qquad\quad 22EI/l \qquad\qquad\quad 1.00$$

$$m_{a1} = -pl^2/12 \quad \text{(counter clockwise bending moment)}$$
$$m_{a2} = +pl^2/12 \quad \text{(clockwise bending moment)}$$

Fig. 3.31 Static scheme of
frame 4–1–2

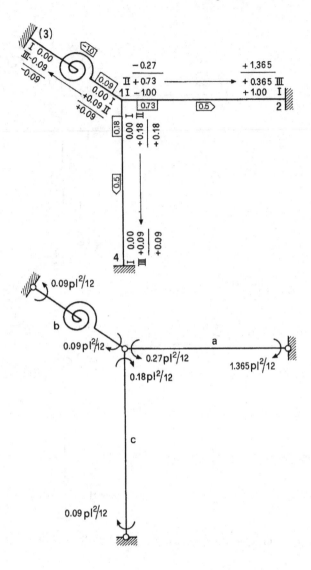

On the sketches of the frames, the subsequent operations of repartition and transmission are written. They lead to the hyperstatic moments listed here below.

– Frame 3–1–4

$$t_{a1} = +0.91\, fl/2$$
$$t_{a2} = +1.09\, fl/2$$
$$m_{b1} = -0.73\, fl/2$$
$$m_{b3} = -0.365\, fl/2$$

$$m_{c1} = -0.18\,f l/2$$
$$m_{c4} = -0.09\,f l/2$$

– Frame 4–1–2

$$m_{a1} = -0.27\,p l^2/12$$
$$m_{a2} = +1.365\,p l^2/12$$
$$t_{b1} = +0.09\,p l^2/12$$
$$t_{b3} = -0.09\,p l^2/12$$
$$m_{c1} = +0.18\,p l^2/12$$
$$m_{c4} = +0.09\,p l^2/12$$

In the examples presented up to here, one had to equilibrate only one joint and this had obtained immediately with one simple moment repartition. When the internal joints of the frames are more than one, the repartition procedure can be applied equilibrating the joints one after the other through iterative cycles repeated until the convergence is reached. This iterative procedure is shown in Fig. 3.32 with reference to the frame analysed at the beginning of the present section (see Fig. 3.8).

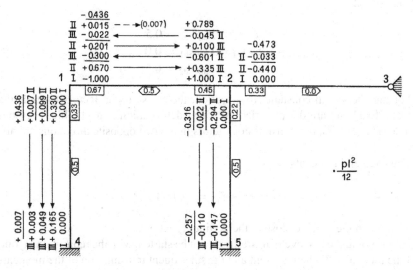

Fig. 3.32 Moment repartition for a plane frame

With $I' = 2I$, one obtains the repartition factors calculated here below.

– Joint 1

$$
\begin{aligned}
\text{``a''} \ 4EI'/l &= 8EI/l : 12EI/l &&= 0.67 \\
\text{``c''} \ 4EI/l &= \underline{4EI/l} : 12EI/l &&= \underline{0.33} \\
&\quad\ \ 12EI/l &&\ \ \ 1.00
\end{aligned}
$$

– Joint 2

$$
\begin{aligned}
\text{``a''} \ 4EI'/l &= 8EI/l : 18EI/l &&= 0.45 \\
\text{``b''} \ 3EI'/l &= 6EI/l : 18EI/l &&= 0.33 \\
\text{``c''} \ 4EI/l &= \underline{4EI/l} : 18EI/l &&= \underline{0.22} \\
&\quad\ \ 18EI/l &&\ \ \ 1.00
\end{aligned}
$$

On the members, the transmission factors are marked:

– from joint 1

$$
\begin{aligned}
\text{``a''} \ t_{21} &= k_{ai}/k_{a2} = 0.5 \\
\text{``c''} \ t_{41} &= k_{ci}/k_{c1} = 0.5
\end{aligned}
$$

– from joint 2

$$
\begin{aligned}
\text{``a''} \ t_{12} &= k_{ai}/k_{a2} = 0.5 \\
\text{``b''} \ t_{32} &= k_{bi}/k_{b1} = 0.0 \\
\text{``d''} \ t_{52} &= k_{di}/k_{d2} = 0.5
\end{aligned}
$$

For members with constant cross section, the transmission factors towards the opposite fixed end are 0.5 provided the shear deformation is negligible. For non-symmetric members, the transmission factors in the two opposite directions are different.

The fixed end moments

$$\overline{m}_{a1} = -\overline{m}_{a2} = -pl_2/12$$

are written on the scheme close to the pertinent sections.

At this point, the first cycle of equilibria can be started, with the repartition at joint 1 of the moments "I" there present and the subsequent transmission of the moments "II" to the opposite ends of the members where the moments "III" in this way arrive, and the repartition at joint 2 of the moments "I" + "III" there present and eventually the transmission of the new moments "II" to the opposite ends of the members.

Again with reference to Fig. 3.32, one can notice that, at the end of this first cycle of equilibria, a non-balanced moment "III" has come back to joint 1. So, the procedure goes on with a new cycle where every time the repartition of the last non-balanced moment "III" of a joint is made and, after the equilibrium in this way obtained with the moments "II", proper lines are marked. At the same time on the other joint, a moment "III" arrives to break the previous equilibrium. So, new cycles of equilibria are repeated in an iterative way until the residual non-balanced moments "III" are small enough. At this point, all the moments are summed up section by section to provide the final solution.

The speedy convergence of this iterative procedure of moment distribution for non-sway frames is due to the prevalence of the direct rotation stiffness with respect to the indirect ones and to the consequent small value of the transmission factors: the deviatory effect that comes back to the joints through the subsequent cycles is every time halved. So, for the frame of Fig. 3.32, after few cycles the non-balanced residual (0.007 marked in brackets) has decreased down to less than 1% of the initial values of the moments. Stopping the procedure at this point, this value of the residual gives the magnitude order of the error left in the solution.

The procedure here described has been presented on 1930 by Hardy Cross, from whom the current denomination of *Cross Method* derives. Its application utility has practically vanished with the development of the electronic computation able to elaborate easily the numerical solution of large sets of equations. The conceptual interest remains connected with the physical interpretation of the moment repartition and transmission operations, with a formative value for the learning of the structural analysis.

3.2.3 Examples of Plane Frames

Some conclusive examples of analysis of plane non-sway frames are presented with reference to the portal structure of Fig. 3.33. In the general case, also if the axial deformations of the members are neglected, the cited structure belongs to the category of sway frames, since the beam can display a horizontal translation ξ. But in this section, the analysis is made on a symmetric arrangement of structure and loads for which the translation ξ remains null.

The unknowns of the problem, following the Displacement Method, are the rotations φ_1, φ_2, φ_3, φ_4 of the four joints, but one can operate with a reduced set of equations due to the equalities

$$\varphi_1 = -\varphi_2$$
$$\varphi_3 = -\varphi_4$$

of the expected symmetric solutions.

So, two equilibrium equations are set for the rotations of the joints 1 and 3:

$$\begin{cases} (m_{11} - m_{12})\varphi_1 + m_{13}\varphi_3 + m_{10} = 0 \\ m_{31}\varphi_1 + m_{33}\varphi_3 + m_{30} = 0 \end{cases}$$

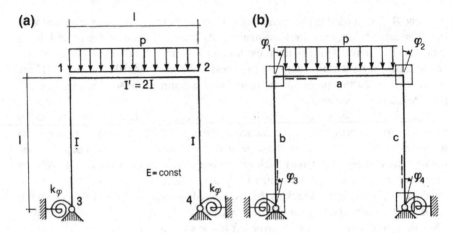

Fig. 3.33 Portal structure with symmetric arrangement

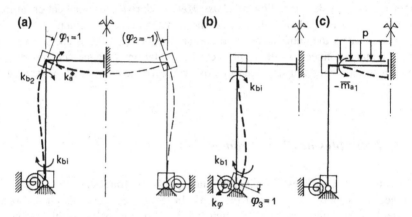

Fig. 3.34 Superposition of effects

In particular on the joint 3 meets, other than the one of the column b, the stiffness k_φ of the rotation spring. This latter, that can simulate the elastic behaviour of the foundation element placed on the ground, quantifies the restoring moment due to a unity rotation of the support, and is expressed in Nm, as inverse quantity of the flexibility c_φ defined at Sect. 2.2.3.

Figure 3.34 describes the superposition of effects, taking into account the symmetry of the structural behaviour. Summing up the contributions at the involved joints 1 and 3, one obtained the equations:

$$\begin{cases} \left(\frac{4EI}{l} + \frac{2EI'}{l}\right)\varphi_1 + \frac{2EI}{l}\varphi_3 - \frac{pl^2}{12} = 0 \\ \frac{2EI}{l}\varphi_1 + \left(k_\varphi + \frac{4EI}{l}\right)\varphi_3 = 0 \end{cases}$$

that, with $I' = 2I$ and with

$$\kappa_\varphi = \frac{k_\varphi}{4EI/l}$$

becomes:

$$\begin{cases} 4\varphi_1 + \varphi_3 = \frac{pl^2}{24EI} \\ \varphi_1 + 2(\kappa_\varphi + 1)\varphi_3 = 0 \end{cases}$$

from which one obtains the solution:

$$\varphi_1 = +\frac{pl^3}{12EI}\frac{\kappa_\varphi + 1}{8\kappa_\varphi + 7}$$

$$\varphi_3 = -\frac{pl^3}{12EI}\frac{1/2}{8\kappa_\varphi + 7}$$

So, one can calculate the moments:

$$M_{a1} = -\frac{pl^2}{12} + \frac{2EI'}{l}\varphi_1 = -\frac{pl^2}{12}\left(1 - 4\frac{\kappa_\varphi + 1}{8\kappa_\varphi + 7}\right) = -\overline{M}c_1$$

$$M_a = M_{a1} + \frac{pl^2}{8} = \overline{M}\left(\frac{3}{2} - c_1\right) = +\overline{M}c_a$$

$$M_{b1} = -\frac{4EI}{l}\varphi_1 - \frac{2EI}{l}\varphi_3 = M_{a1}$$

$$M_{b3} = \frac{2EI}{l}\varphi_1 + \frac{4EI}{l}\varphi_3 = +\frac{pl^2}{12}\frac{2\kappa_\varphi}{8\kappa_\varphi + 7} = +\overline{M}c_3$$

$$(M_{b3} = -k_\varphi\varphi_3)$$

Figure 3.35 gives the diagrams of the internal forces for the two limit situations, one for rotatory restraints very weak ($k_\varphi \to 0$) as if the columns were hinged at their base, the other for rotatory restraints very strong ($k_\varphi \to \infty$) as if the columns had full supports at their base, and also for an intermediate base support situation with a rotation stiffness corresponding to $\kappa_\varphi = 1$. For the bending moments, as a ratio to the "reference" one $\overline{M} = pl^2/12$, one has the values obtainable from the following table.

k_φ	c_a	c_1	c_3	c_o
0	15/14	3/7	0	1/14
1	31/30	7/15	2/15	1/10
∞	1/1	1/2	1/4	1/8

One can notice how the rotation stiffness of the base supports has a small influence on the bending moments of the beam, with variations between the situations of hinged supports and fully restrained supports contained within 7% of the "reference" moment $\overline{M}$.

Eventually, the shear and axial forces are deduced from the following equilibria:

$$V'_a = -V''_a = +\frac{pl}{2} = +\overline{V}$$

$$V_b = -V_c = \frac{M_{b1} - M_{b3}}{1} = -\frac{pl}{2}\frac{1}{2}\frac{2\kappa_\varphi + 1}{8\kappa_\varphi + 7} = -\overline{V}c_o$$

$$N_b = N_c = -\frac{pl}{2} = -\overline{V}$$

$$N_a = +V_b = -\overline{V}c_o$$

where the values of c_o can be deduced from the preceding table.

For the same portal analysed above, with reference only to the case of columns fully restrained at their base, the analysis is now elaborated for the two conditions

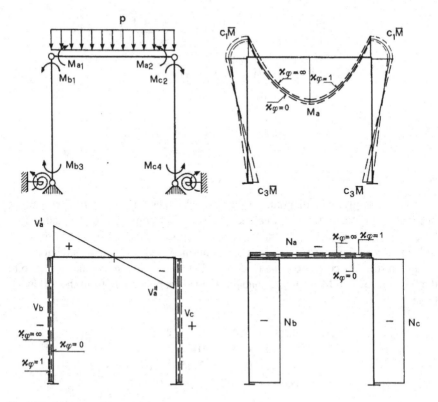

Fig. 3.35 Diagrams of internal forces

of symmetric thermal action described in Fig. 3.36. The first condition consists of a uniform variation Δt_o of temperature referred to the axis of the beam; the second condition consists of a differential variation $\pm \Delta t$ of temperature between the lower and upper sides of the same beam.

With $\kappa_\varphi = \infty$, the set of equilibrium conditions is reduced to only one equation:

$$\frac{8EI}{1}\varphi_1 + m_{10} = 0$$

that leads to the solution:

$$\varphi_1 = -\frac{1}{8EI}m_{10} \quad (-\varphi_2)$$

$$(\varphi_3 \equiv 0 \equiv \varphi_4)$$

For the first condition of uniform variation of temperature (Fig. 3.36a), one has at joint 1 the fixed end moment coming from the column:

$$m_{10} = \overline{m}_{b2} = +\frac{6EI}{1^2}\delta$$

with

$$\delta = \alpha_t \Delta t_o \frac{1}{2}$$

So, the joint rotation is

$$\varphi_1 = -\frac{3}{8}\alpha_t \Delta t_o$$

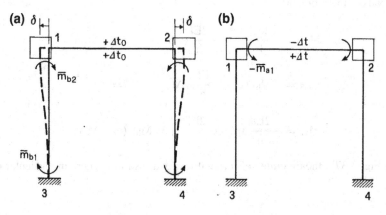

Fig. 3.36 Portal under thermal actions

and leads to the moments:

$$M_{a1} = +\frac{2EI'}{l}\varphi_1 = -\frac{3}{2}\frac{EI}{l}\alpha_t\Delta t_o \qquad\qquad (= +M_{a2})$$

$$M_{b1} = -\frac{3EI}{l}\alpha_t\Delta t_o - \frac{4EI}{l}\varphi_1 = M_{a1} \qquad\qquad (= +M_{c2})$$

$$M_{b3} = +\frac{3EI}{l}\alpha_t\Delta t_o + \frac{2EI}{l}\varphi_1 = +\frac{9}{4}\frac{EI}{l}\alpha_t\Delta t_o \ (= +M_{c4})$$

In Fig. 3.37a, the diagrams of internal forces are shown where in particular one has:

$$V_a = 0$$
$$V_b = \frac{M_{b1} - M_{b3}}{l} = -\frac{15}{4}\frac{EI}{l^2}\alpha_t\Delta t_o \ (= -V_c)$$
$$N_a = V_b$$
$$N_b = 0 \qquad\qquad\qquad\qquad (= +N_c)$$

For the second condition of a differential temperature variation (Fig. 3.36b), one has at joint 1 the fixed end moment coming from the beam (see Fig. 2.93):

$$m_{10} = \overline{m}_{a1} = -\frac{2EI'}{h}\alpha_t\Delta t$$

with h depth of the section. So, the rotation at joint 1 is:

$$\varphi_1 = +\frac{1}{2}\frac{1}{h}\alpha_t\Delta t$$

and leads to the moments

$$M_{b1} = -\frac{4EI}{l}\varphi_1 = -\frac{2EI}{h}\alpha_t\Delta t \qquad\qquad (= +M_{c2})$$

$$M_{b3} = -M_{b1}/2 = +\frac{EI}{h}\alpha_t\Delta t \qquad\qquad (= +M_{c4})$$

$$M_{a1} = -\frac{2EI'}{h}\alpha_t\Delta t + \frac{2EI'}{l}\varphi_1 = M_{b1} \ (= +M_{a2})$$

In Fig. 3.37b, the diagrams of internal forces are shown where in particular one has:

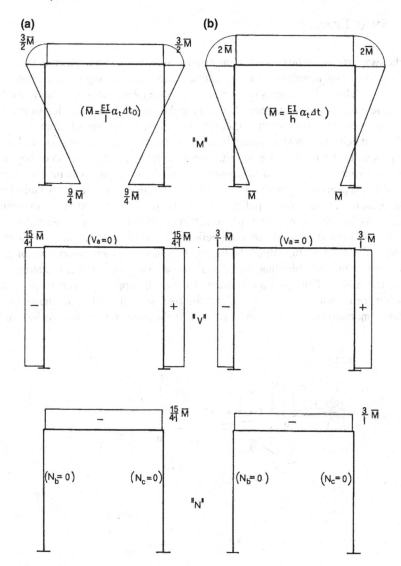

Fig. 3.37 Diagrams of internal forces

$$V_a = 0$$
$$V_b = \frac{M_{b1} - M_{b3}}{l} = -3\frac{EI}{lh}\alpha_t \Delta t \ (= -V_c)$$
$$N_a = V_b$$
$$N_b = 0 \qquad\qquad\qquad (= +N_c)$$

3.3 Sway Frames

For the sway frame, on the basis of what stated at Sect. 2.3.3, also neglecting the axial deformations of the members some translation components of the joint displacements remain free. These components are to be added to the joint rotations in the list of the geometric unknowns, and in general, assuming the quoted approximation of axially non-deformable members, each of them can be referred to groups of joints.

This will be fully shown in the application examples of the present section. These examples will refer to plane frames and mainly to those arranged on an orthogonal mesh of vertical columns and horizontal beams. A short note will be eventually devoted to the analysis of the three-dimensional structures of multi-storey buildings.

Beforehand, an example of application of the Displacement Method is shown with reference to the analysis of a grid structure, that is a system of beams assembled in a horizontal plane and subjected to loads acting in the vertical direction, as those that constitute the floors of the buildings. Figure 3.38 shows an elementary case of grid with only one internal joint that can be analysed with a small number of unknowns.

Joints 1 and 4 of the grid are restrained with full supports, while joints 3 and 5 are supported with bearings that allow the free flexural rotation of the beam in its direction, restraining the torsional rotation in the orthogonal plane. As indicated

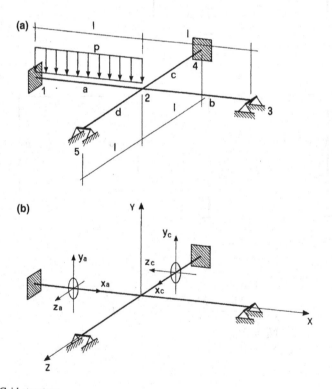

Fig. 3.38 Grid structure

in Fig. 3.38b, the structure is referred to a system of axes X, Y, Z with the axis Y orthogonal to the grid plane and the other two parallel to the sides of the structural mesh. The local systems of axes x, y, z of the beams are oriented with the axis y parallel to the axis Y of the global system so to have vertical planes of uniaxial flexure. The orthogonality of the two sets of beams avoids the necessity of projections of moments and rotations similar to those presented at Sect. 3.1.3 for the forces and translations of plane frames.

The auxiliary structure is obtained placing on joint 2 an additional support with three degrees of restraint correspondent to the original displacement freedoms of the joint for the grid behaviour examined: the first restraint fixes the vertical translation η, and the other two fix the rotations φ_X, φ_Z around the horizontal axes.

Figure 3.39 shows the superposition of effects. Using the symbols a, b, c, d as member references, the symbols k, k', k_v, k'_v and k_t, respectively, for the direct and indirect rotation stiffness, for the direct and indirect translation stiffness and for the torsional stiffness, and the symbols $\bar{v}'_a$, $\bar{v}''_a$, $\bar{m}'_a$, $\bar{m}''_a$ for fixed support forces and moments at the two ends of the loaded beam, one has the equations:

$$\begin{cases} (k_{av} + k_{bv} + k_{cv} + k_{dv})\eta - (k'_{cv} + k'_{dv})\varphi_X + (k'_{av} + k'_{bv})\varphi_Z + \bar{v}''_a = 0 \\ -(k'_{cv} + k'_{dv})\eta + (k_{at} + k_{bt} + k_c + k_d)\varphi_X = 0 \\ (k'_{av} + k'_{bv})\eta + (k_a + k_b + k_{ct} + k_{dt})\varphi_Z + \bar{m}''_a = 0 \end{cases}$$

where the first equation expresses the vertical translation equilibrium of joint 2, the second equation expresses the rotation equilibrium around X of the same joint and the third equation expresses its rotation equilibrium around Z.

The solution is given in the case of all equal constant section beams with a ratio between the torsional and the flexural stiffness equal to

$$\frac{GJ}{EI} = \frac{1}{2}$$

One has

$$k_a = k_c = \frac{4EI}{1} \qquad \left(k'_a = k'_c = \frac{2EI}{1} \right)$$

$$k_b = k_d = \frac{3EI}{1} \qquad (k'_b = k'_d = 0)$$

$$k'_{av} = k'_{cv} = \frac{6EI}{1^2}$$

$$k'_{bv} = k'_{dv} = \frac{3EI}{1^2}$$

Fig. 3.39 Superposition of effects

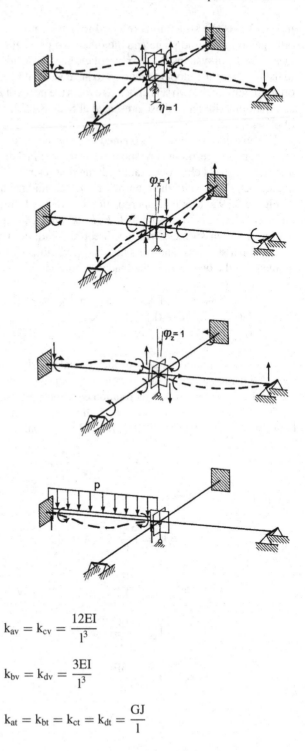

$$k_{av} = k_{cv} = \frac{12EI}{l^3}$$

$$k_{bv} = k_{dv} = \frac{3EI}{l^3}$$

$$k_{at} = k_{bt} = k_{ct} = k_{dt} = \frac{GJ}{l}$$

With the fixed end forces and moments:

$$\bar{v}_a'' = +\frac{pl}{2} \quad (\bar{v}_a' = +\frac{pl}{2})$$

$$\bar{m}_a'' = +\frac{pl^2}{12} \quad (\bar{m}_a' = -\frac{pl^2}{12})$$

the equations become

$$\begin{cases} \dfrac{30}{1}\eta - 3\varphi_x + 3\varphi_z = -\dfrac{pl^3}{2EI} \\[2mm] -\dfrac{3}{1}\eta + 8\varphi_x = 0 \\[2mm] +\dfrac{3}{1}\eta + 8\varphi_z = -\dfrac{pl^3}{12EI} \end{cases}$$

and lead to:

$$\eta = -0.01689\frac{pl^4}{EI}$$

$$\varphi_x = -0.00633\frac{pl^3}{EI}$$

$$\varphi_z = -0.00408\frac{pl^3}{EI}$$

Setting $\overline{M} = pl^2/12$, one obtains the following moments:

$$M_{a1} = -\overline{M} + \frac{6EI}{1^2}\eta + \frac{2EI}{1}\varphi_z = -2.314\overline{M}$$

$$M_{a2} = -\overline{M} - \frac{6EI}{1^2}\eta - \frac{4EI}{1}\varphi_z = +0.412\overline{M}$$

$$M_a = \frac{3}{2}\overline{M} + (M_{a1} + M_{a2})/2 = +0.549\overline{M}$$

$$T_a = \frac{GJ}{1}\varphi_x = -0.038\overline{M} = -T_b$$

$$M_{b2} = -\frac{3EI}{1^2}\eta + \frac{3EI}{1}\varphi_z = +0.461\overline{M}$$

$$M_{c4} = +\frac{6EI}{1^2}\eta - \frac{2EI}{1}\varphi_x = -1.064\overline{M}$$

$$M_{c2} = -\frac{6EI}{1^2}\eta + \frac{4EI}{1}\varphi_x = +0.912\overline{M}$$

$$T_c = \frac{GJ}{1}\varphi_z = -0.024\overline{M} = -T_d$$

$$M_{d2} = -\frac{3EI}{l^2}\eta - \frac{3EI}{l}\varphi_X = +0.836\overline{M}$$

where the bending moments are assumed positive when the lower side is tensioned and the torsional moments are assumed positive when they correspond to a clockwise screwing.

Figure 3.40 shows the diagrams of the internal forces where in particular (with $\overline{V} = pl/2$) the shear forces are:

$$V_{a1} = +\overline{V} + \frac{M_{a2} - M_{a1}}{l} = +1.454\overline{V}$$

$$V_{a2} = -\overline{V} + \frac{M_{a2} - M_{a1}}{l} = -0.546\overline{V}$$

$$V_b = -\frac{M_{b2}}{l} = -0.077\overline{V}$$

$$V_c = -\frac{M_{c4} - M_{c2}}{l} = -0.329\overline{V}$$

$$V_d = +\frac{M_{d2}}{l} = +0.139\overline{V}$$

expressed with the conventions shown in the figure.

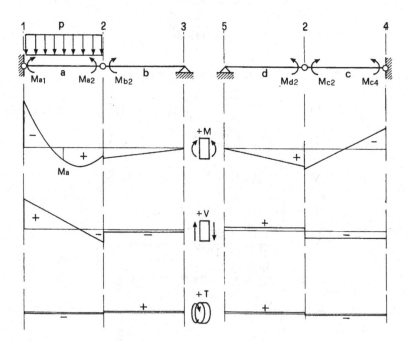

Fig. 3.40 Diagrams of internal forces

To control the solution, one can verify the translation and rotation equilibria of joint 2 with:

$$-V_{a2} + V_b + V_c - V_d = 0$$
$$+M_{c2} - M_{d2} + T_a - T_b = 0$$
$$-M_{a2} + M_{b2} + T_c - T_d = 0$$

For the case examined, one can notice, other than the small contribution of the torsional moments, the small bearing contribution given to the loaded beam 1–2–3 by the transverse beam 4–2–5 due to its low stiffness. So, this latter does not represent a vertical restraint like that given by the end supports of the loaded beam like a column could give, but constitutes the second flexural element of similar effectiveness on which the loaded beam can transmit part of the action. The share of transmitted action is related to the ratio between the stiffness of the two orthogonal beams.

3.3.1 Force Repartition on Multi-Portal Frames

The behaviour of the widespread portal frame typology is now examined with reference to the horizontal actions. In the preceding Sect. 2.3.3, the behaviour of a simple one-bay portal has already been analysed (see Fig. 2.65) pointing out the influence of the ratio between the flexural features of beam and columns with reference to the response to a horizontal force applied to the beam.

The analysis is extended now to the multi-bay portals, beginning with the elementary case of beams all supported with hinges at the top of the columns (Fig. 3.41a), following a typical arrangement of the precast structures. This portal structure, subjected to a horizontal top force H, is solved, on the assumption of negligible axial deformations of the beams, with only one geometric unknown corresponding to the horizontal translation ξ of the additional support that restrains, in the auxiliary structure, the beams' line (Fig. 3.41b).

Figure 3.42 shows the superposition of effects with which the terms of the solving equation are evaluated following the Displacement Method. This equation expresses the translation equilibrium of the beams' line:

$$r_{xx}\,\xi + r_{x0} = 0$$

giving a null value to the reaction of the additional support.

Calling k_{vj} the direct translation stiffness of the generic column j ($=3EI_j/h^3$ for columns with constant cross section and height h), the coefficient r_{xx} is equal to the sum of such stiffness of the n columns and, being $r_{x0} = -H$, the solution is

$$\xi = \frac{H}{\Sigma k_{vj}}$$

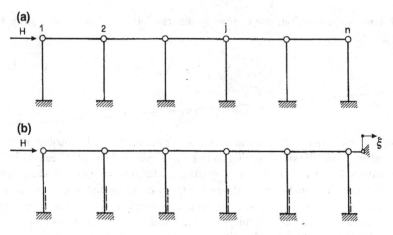

Fig. 3.41 Multi-bay hinged portal

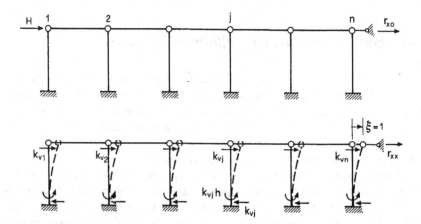

Fig. 3.42 Superimposition of effects

This solution shows a common collaboration of the columns that leads to a translation inversely proportional to the global stiffness of the structural system. With the computed translation, one deduces for any column the share of horizontal force applied to it:

$$V_j = k_{vj}\xi = \frac{k_{vj}}{\Sigma k_{vj}}H$$

This equation shows a repartition of the global force H proportional to the stiffness of each column. Any single force V_j corresponds to the (constant) shear of the pertinent column and, multiplied by the height h, gives the bending moment at its base.

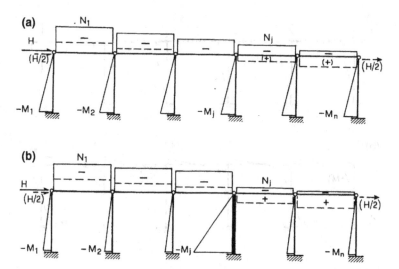

Fig. 3.43 Diagrams of internal forces

The diagrams of the bending moment on the columns are drawn in Fig. 3.43a in the case of all equal stiffnesses and in Fig. 3.43b in the case of a column j much stiffer than the others (e.g. $I_j = 5I$). In the first case, the base moments are calculated simply with $M = -Hh/n$. In the second case, the base moment of the stiffer column would be $M = -5Hh/(n + 4)$ five times bigger than those of the other columns.

Eventually on the beams, the diagrams of the axial forces are drawn; through these forces, the applied action H is distributed on the single columns. These axial forces are calculated subtracting progressively, starting from the end where the action is applied, the shear forces coming from the subsequent columns. The quoted diagrams depend on the actual position of the action: if this action was applied half on the joint 1 and half on the joint n, instead of entirely on the joint 1, one would have the dashed diagrams drawn in the same Fig. 3.43.

If beams and columns were jointed by moment-resisting connections, the solution would depend also on the rotations of the joints. So, these rotations φ_j shall be added as unknowns to the translation ξ of the beams' line.

First, the limit case of beams with infinite stiffness is examined. No sensible rotations would arise in this case, as shown in Fig. 3.44a for the concerned multi-bay portal structure that, with $\varphi_1 \equiv \varphi_2 \equiv \cdots \equiv \varphi_j \equiv \cdots \varphi_n \equiv 0$, is analysed in the same way of the preceding hinged structure. In the same equilibrium equation, only the values of the translation stiffness are to be modified with reference to the new end support placed at the top of the columns e.g. $k_{vj} = 12EI/h^3$ in the case of constant cross section.

For the same applied action H and for the same column section, one obtains a translation ξ much smaller than for the hinged structure: 4 times smaller in the case of constant cross section like the inverse ratio 3/12 between the respective stiffness.

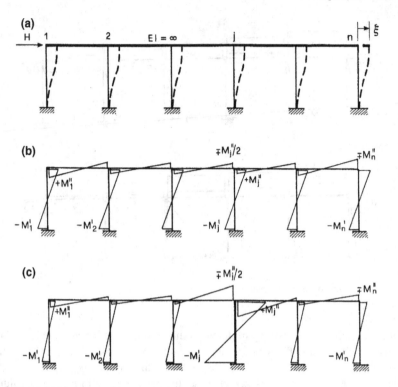

Fig. 3.44 Multi-bay portal with moment-resisting connections

With the computed translation, one deduces for any column the share of horizontal force applied to it, which is again proportional to its stiffness. Any single force V_j corresponds to the (constant) shear of the pertinent column. With the indirect translation stiffness, one calculates eventually the bending moments at the base and top of the columns. For a constant cross section, being

$$k_{v1} = k_{v2} = \frac{6EI}{h^2} = k_v \frac{h}{2}$$

one has at the base and top the same bending moments equal to:

$$M_j' = -M_j'' = V_j \frac{h}{2}$$

So, the big flexural stiffness of the beams leads to halving the maximum bending moment, subdividing the global variation $V_j h$ between the base and the top.

Figure 3.44b, c show the diagrams of the bending moment for the two cases similar to the preceding ones: for all equal columns for which the maximum moments are calculated simply with $Hh/(2n)$; with a column much stiffer than the others for which the moment is increased proportionally. To be noted that the bending moments, through the full joint restraints, are transmitted also to the beams: on the two end joints 1 and n, the whole bending moment coming from the column goes to the connected beam; on the internal joints, it is subdivided between the two connected beams.

Multi-bay portal structures with moment-resisting joint connections have actually an intermediate behaviour, with $V_j h/2 < M'_j < V_j h$ and $V_j h/2 > M''_j > 0$: closer to that of hinged connections if the flexural stiffness of the beam is small and closer to that of the opposite situation if the flexural stiffness of the beam is big with reference to the column stiffness. The different rotations of the joints modify also the repartition of the horizontal forces on the columns that cannot be expressed with simple proportionalities as in the preceding cases.

Let's consider, for example, the symmetric frame analysed on the halved scheme of Fig. 3.45 considering the anti-symmetric arrangement of the load. Following the Displacement Method and with the usual assumptions, one has the four geometric unknowns φ_1, φ_2, φ_3 and ξ as shown in Fig. 3.45b. For all equal columns with constant cross section of stiffness EI and for all equal beams with constant cross section of stiffness EI', one has the following equations:

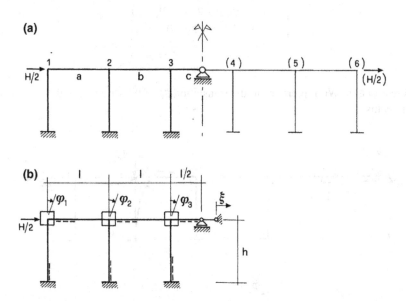

Fig. 3.45 Analysis of symmetric frame

$$
\begin{cases}
\left(\dfrac{4EI}{h}+\dfrac{4EI'}{l}\right)\varphi_1+\dfrac{2EI'}{l}\varphi_2-\dfrac{6EI}{h^2}\xi=0 \\[4mm]
\dfrac{2EI'}{l}\varphi_1+\left(\dfrac{4EI'}{l}+\dfrac{4EI}{h}+\dfrac{4EI'}{l}\right)\varphi_2+\dfrac{2EI'}{l}\varphi_3-\dfrac{6EI}{h^2}\xi=0 \\[4mm]
\dfrac{2EI'}{l}\varphi_2+\left(\dfrac{4EI'}{l}+\dfrac{4EI}{h}+\dfrac{3EI'}{1/2}\right)\varphi_3-\dfrac{6EI}{h^2}\xi=0 \\[4mm]
-\dfrac{6EI}{h^2}\varphi_1-\dfrac{6EI}{h^2}\varphi_2-\dfrac{6EI}{h^2}\varphi_3+3\dfrac{12EI}{h^3}\xi=\dfrac{H}{2}
\end{cases}
$$

written with the usual superposition of effects (Fig. 3.46). The first three equations express the rotation equilibria of joints 1, 2 and 3, while the last equation expresses the horizontal translation equilibrium of the beams' line.

With $l=h$ and $I'=2I$, the equations become

$$
\begin{cases}
6\varphi_1+2\varphi_2-3\bar\xi=0 \\
2\varphi_1+10\varphi_2+2\varphi_3-3\bar\xi=0 \\
2\varphi_2+12\varphi_3-3\bar\xi=0 \\
-\varphi_1-\varphi_2-\varphi_3+6\bar\xi=\dfrac{Hh^2}{12EI}
\end{cases}
$$

where $\bar\xi=\xi/h$. With reference to the translation ξ_o of the structure with infinitively stiff beams

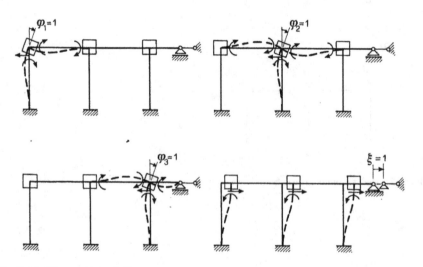

Fig. 3.46 Superposition of effects

$$\bar{\xi}_o = \frac{\xi_o}{h} = \frac{Hh^2}{72EI}$$

one obtains the solution:

$$\varphi_1 = +\frac{16}{31}\bar{\xi}_o = +0.516\bar{\xi}_o$$

$$\varphi_2 = +\frac{6}{31}\bar{\xi}_o = +0.194\bar{\xi}_o$$

$$\varphi_3 = +\frac{8}{31}\bar{\xi}_o = +0.258\bar{\xi}_o$$

$$\bar{\xi} = +\frac{36}{31}\bar{\xi}_o = +1.161\bar{\xi}_o$$

showing that the rotation flexibility of the upper joints of the portal leads, in the case here examined, to an increased translation of the top beams' line by about 16%.

In the following, the base and top bending moments of the columns are calculated (with $\overline{M} = Hh/12$).

$$M_1' = +\frac{2EI}{h}\varphi_1 - \frac{6EI}{h^2}\xi = -\frac{92}{93}\overline{M} = -0.989\overline{M}$$

$$M_1'' = -\frac{4EI}{h}\varphi_1 + \frac{6EI}{h^2}\xi = +\frac{76}{93}\overline{M} = +0.817\overline{M}$$

$$M_2' = +\frac{2EI}{h}\varphi_2 - \frac{6EI}{h^2}\xi = -\frac{102}{93}\overline{M} = -1.097\overline{M}$$

$$M_2'' = -\frac{4EI}{h}\varphi_2 + \frac{6EI}{h^2}\xi = +\frac{96}{93}\overline{M} = +1.032\overline{M}$$

$$M_3' = +\frac{2EI}{h}\varphi_3 - \frac{6EI}{h^2}\xi = -\frac{100}{93}\overline{M} = -1.075\overline{M}$$

$$M_3'' = -\frac{4EI}{h}\varphi_3 + \frac{6EI}{h^2}\xi = +\frac{92}{93}\overline{M} = -0.989\overline{M}$$

The corresponding shear forces are (with $\overline{V} = H/6$):

$$V_1 = (M_1'' - M_1')/h = +\frac{84}{93}\overline{V} = +0.903\overline{V}$$

$$V_2 = (M_2'' - M_2')/h = +\frac{99}{93}\overline{V} = +1.065\overline{V}$$

$$V_3 = (M_3'' - M_3')/h = +\frac{96}{93}\overline{V} = +1.032\overline{V}$$

These values show how a good stiffness of the beams (with $EI'/l = 2EI/h$) leads to almost uniform distribution of the forces on the columns, with bending moments that differ less than 10% from those obtained with infinitive stiff beams.

The bending moments and shear forces on the beams are calculated with:

$$M_{a1} = +\frac{4EI'}{1}\varphi_1 + \frac{2EI'}{1}\varphi_2 = +\frac{76}{93}\overline{M} = +0.817\overline{M}$$

$$M_{a2} = -\frac{2EI'}{1}\varphi_1 - \frac{4EI'}{1}\varphi_2 = -\frac{56}{93}\overline{M} = -0.602\overline{M}$$

$$M_{b2} = +\frac{4EI'}{1}\varphi_2 + \frac{2EI'}{1}\varphi_3 = +\frac{40}{93}\overline{M} = +0.430\overline{M}$$

$$M_{b3} = -\frac{2EI'}{1}\varphi_2 - \frac{4EI'}{1}\varphi_3 = -\frac{44}{93}\overline{M} = -0.473\overline{M}$$

$$M_{c3} = +\frac{3EI'}{1/2}\varphi_3 = +\frac{48}{93}\overline{M} = +0.516\overline{M}$$

$$V_a = (M_{a2} - M_{a1})/1 = -\frac{66}{93}\overline{V} = -0.710\overline{V}$$

$$V_b = (M_{b3} - M_{b2})/1 = -\frac{42}{93}\overline{V} = -0.452\overline{V}$$

$$V_c = -2M_{c3}/1 = -\frac{48}{93}\overline{V} = -0.516\overline{V}$$

And eventually, assuming that the force H is applied half to the joint 1 and half to the joint 6 of the frame, one has:

$$N_a = -\frac{H}{2} + V_1 = -2.097\overline{V}$$

$$N_b = N_a + V_2 = -1.032\overline{V}$$

$$N_c = N_b + V_3 = 0.000\overline{V}$$

$$N_1 = -V_a = +0.710\overline{V}$$

$$N_2 = +V_a - V_b = -0.258\overline{V}$$

$$N_3 = +V_b - V_c = +0.064\overline{V}$$

In Fig. 3.47, the diagrams of the internal forces calculated above can be found. The limits of the rigid beams are marked with thin lines for the pertinent comparisons. For what concerns the conventions, as usual the bending moment diagrams are placed at the tensioned side, and the clockwise shear and the tensile axial forces are taken as positive.

3.3.2 Multi-storey Frames

Among the sway frames, a wide class is constituted by the structures arranged in an orthogonal assembly of vertical columns and horizontal beams. The *multi-storey*

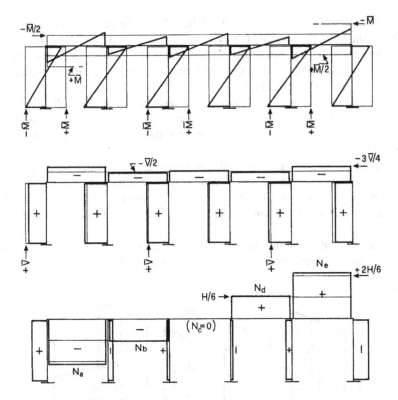

Fig. 3.47 Diagrams of internal forces

frames belong to this class: they form the principal bearing skeleton of many buildings.

These frames are inserted in three-dimensional structural assemblies often of large dimensions. Also adopting the simplifications allowed by the specific features of the examined typologies, the stress analysis of these structural assemblies requires the elaboration of very large sets of equations, unless it can be subdivided into proper partial schemes of smaller dimensions.

Let's consider, for example, the types of tall buildings presented in Fig. 3.48. The bearing skeleton of these buildings includes horizontal decks constituted by floors and beams. The decks provide to the structural assembly a number of diaphragms which are very rigid in their plane and able to distribute the horizontal actions on the different vertical frames.

With reference to the complexity of the calculations and the possible simplifications, two situations have to be distinguished: the one of *braced buildings* and the one of *unbraced buildings*. In the first situation, the resistance to the horizontal actions is assigned to special additional elements; in the second situation, the resistance to the horizontal actions is left to the current frame systems of beams and columns.

The bearing skeleton of the braced buildings has some vertical elements with a stiffness much higher than that of the columns. In the preceding section, it has been shown how such coupling brings, due to the compatibility of the floor displacements, most of the horizontal actions on the stiffer element, leaving sensibly unloaded the others (see Figs. 3.43b and 3.44c).

The bracing elements of the buildings can be provided by the *wall core* of the staircase (Fig. 3.48b), by special *shear walls* distributed in the two principal directions (Fig. 3.48c) or by *truss works* such as those widely used in the steel structures (Fig. 3.48d).

The presence of bracing elements that prevent relevant horizontal displacements of the floors allows to treat the single vertical frames of the structural assembly as non-sway frames and to analyse them only for the pertinent vertical loads. Proper approximate partial static schemes could be also extracted, such as those shown at Sect. 3.4.1.

For the verifications of the *overall stability* of the structural assembly against the horizontal actions, the action repartition on the bracing element shall be analysed.

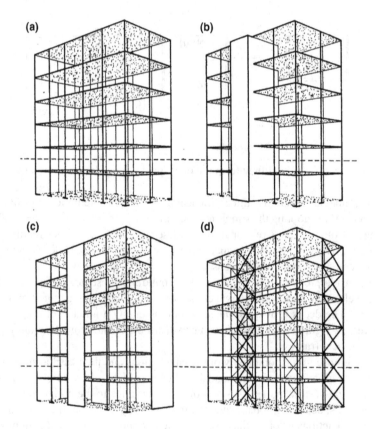

Fig. 3.48 Different types of bearing skeletons for tall buildings

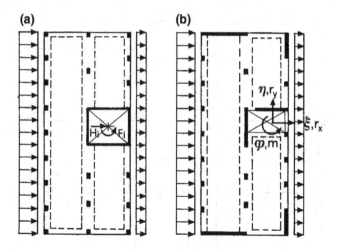

Fig. 3.49 Repartition of floor actions on the bracing elements. **a** Wall core and **b** shear walls

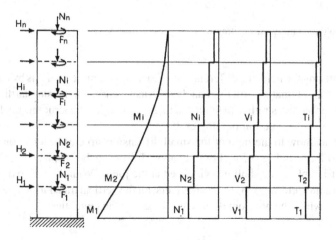

Fig. 3.50 Internal forces in the wall core

In the general case, this analysis requires three equilibrium equations for any floor (Fig. 3.49b), but it can be reduced to a simple calculation of resultants (Fig. 3.49a). The design of the single bracing elements, assumed as vertical cantilever members fixed at their base, is then performed (Fig. 3.50).

The analysis of the multi-storey frame structures of the unbraced buildings leads to the elaboration of very large sets of equations as already said. Following the standard procedure of the Displacement Method, there are six unknowns for every joint. Also introducing, in addition to those related to the negligible axial deformations of the members, the relevant reductions coming from the floor diaphragm links, the dimensions of the numerical algorithms exceed the limit assumed in the present

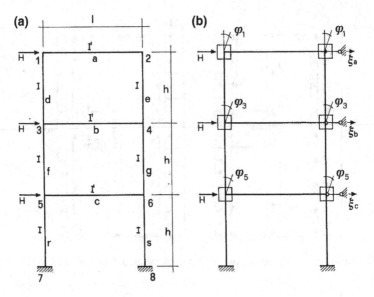

Fig. 3.51 Solution of three-storey frame

volume, entering the field of a different discipline, the structural analysis by electronic computers (or automated structural analysis), that focuses on the standardization of the solving procedures rather than on the interpretation of the structural behaviours of the principal construction typologies.

Going back now to the field of the small dimension structures, let's consider the three-storey frame of Fig. 3.51a. For its analysis following the Displacement Method, the geometric unknowns shall be evidenced at the joints. With the approximation to neglect the axial deformations of the members, all the vertical translation components remain null, while the horizontal ones are linked by the equalities:

$$\xi_1 = \xi_2 = \xi_a$$
$$\xi_3 = \xi_4 = \xi_b$$
$$\xi_5 = \xi_6 = \xi_c$$

With reference to the anti-symmetric load condition shown in the figure, one has again the equalities:

$$\varphi_1 = \varphi_2$$
$$\varphi_3 = \varphi_4$$
$$\varphi_5 = \varphi_6$$

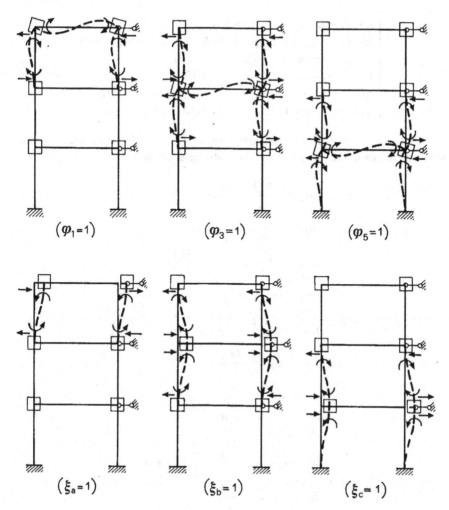

Fig. 3.52 Superimposition of effects

and so the frame can be analysed on the basis of the six geometric unknowns marked in Fig. 3.51b.

In Fig. 3.52, the superposition of effects is shown, leading to the set of six equations written below, referred three to the rotation equilibria of the joints 1, 3, 5 and three to the translation equilibria of the beams a, b, c. The symbols r_a, r_b, r_c refer to half of the pertinent reactions of the additional translation supports of the auxiliary structure.

$$
\begin{cases}
m_{11}\varphi_1 + m_{13}\varphi_3 + m_{1a}\xi_a + m_{1b}\xi_b = 0 \\
m_{31}\varphi_1 + m_{33}\varphi_3 + m_{35}\varphi_5 + m_{3a}\xi_a + m_{3b}\xi_b + m_{3c}\xi_c = 0 \\
m_{53}\varphi_3 + m_{55}\varphi_5 + m_{5b}\xi_b + m_{5c}\xi_c = 0 \\
r_{a1}\varphi_1 + r_{a3}\varphi_3 + r_{aa}\xi_a + r_{ab}\xi_b = H/2 \\
r_{b1}\varphi_1 + r_{b3}\varphi_3 + r_{b5}\varphi_5 + r_{ba}\xi_a + r_{bb}\xi_b + r_{bc}\xi_c = H/2 \\
r_{c1}\varphi_3 + r_{c3}\varphi_5 + r_{cb}\xi_b + r_{cc}\xi_c = H/2
\end{cases}
$$

In the case of members with constant cross section, referring to the usual expressions of the flexural rotatory and translator stiffness, one has:

$$
m_{11} = +\frac{6EI'}{l} + \frac{4EI}{h}
$$

$$
m_{31} = m_{13} = +\frac{2EI}{h}
$$

$$
m_{33} = +\frac{4EI}{h} + \frac{6EI'}{l} + \frac{4EI}{h}
$$

$$
m_{53} = m_{35} = +\frac{2EI}{h}
$$

$$
m_{55} = +\frac{4EI}{h} + \frac{6EI'}{l} + \frac{4EI}{h}
$$

$$
r_{a1} = m_{1a} = -\frac{6EI}{h^2}
$$

$$
r_{a3} = m_{3a} = -\frac{6EI}{h^2}
$$

$$
r_{b1} = m_{1b} = +\frac{6EI}{h^2}
$$

$$
r_{b3} = m_{3b} = +\frac{6EI}{h^2} - \frac{6EI}{h^2} \quad (= 0)
$$

$$
r_{b5} = m_{5b} = -\frac{6EI}{h^2}
$$

$$
r_{c3} = m_{3c} = +\frac{6EI}{h^2}
$$

$$
r_{c5} = m_{5c} = +\frac{6EI}{h^2} - \frac{6EI}{h^2} \quad (= 0)
$$

$$
r_{aa} = +\frac{12EI}{h^3}
$$

$$
r_{ba} = r_{ab} = -\frac{12EI}{h^3}
$$

$$
r_{bb} = +\frac{12EI}{h^3} + \frac{12EI}{h^3}
$$

$$
r_{cb} = r_{bc} = -\frac{12EI}{h^3}
$$

$$r_{cc} = +\frac{12EI}{h^3} + \frac{12EI}{h^3}$$

With $k' = EI'/l = EI/h = k$ and setting $\bar{\xi}_0 = Hh^2/24EI = \xi_0/h$, the equations become:

$$\begin{cases} +5\varphi_1 + \varphi_3 - 3\bar{\xi}_a + 3\bar{\xi}_b = 0 \\ +\varphi_1 + 7\varphi_3 + \varphi_5 - 3\bar{\xi}_a + 3\bar{\xi}_c = 0 \\ +\varphi_3 + 7\varphi_5 - 3\bar{\xi}_b = 0 \\ -\varphi_1 - \varphi_3 + 2\bar{\xi}_a - 2\bar{\xi}_b = 2\bar{\xi}_0 \\ +\varphi_1 - \varphi_5 - 2\bar{\xi}_a + 4\bar{\xi}_b - 2\bar{\xi}_c = 2\bar{\xi}_0 \\ +\varphi_3 - 2\bar{\xi}_b + 4\bar{\xi}_c = 2\bar{\xi}_0 \end{cases}$$

where the unknowns have been homogenized with $\bar{\xi}_a = \xi_a/h, \bar{\xi}_b = \xi_b/h, \bar{\xi}_c = \xi_c/h$. From the first three equations, the rotations can be expressed as a function of the translations with:

$$233\varphi_1 = +123\bar{\xi}_a - 141\bar{\xi}_b + 21\bar{\xi}_c$$
$$233\varphi_3 = +84\bar{\xi}_a + 6\bar{\xi}_b - 105\bar{\xi}_c$$
$$233\varphi_5 = -12\bar{\xi}_a + 99\bar{\xi}_b + 15\bar{\xi}_c$$

Substituting the expressions in the last three equations, one obtains:

$$\begin{cases} +0.556\bar{\xi}_a - 0.710\bar{\xi}_b + 0.180\bar{\xi}_c = \bar{\xi}_0 \\ -0.710\bar{\xi}_a + 1.485\bar{\xi}_b - 0.987\bar{\xi}_c = \bar{\xi}_0 \\ +0.180\bar{\xi}_a - 0.9875\bar{\xi}_b + 1.775\bar{\xi}_c = \bar{\xi}_0 \end{cases}$$

that gives the translations:

$$\xi_a = +13.70\,\xi_0$$
$$\xi_b = +10.60\,\xi_0$$
$$\xi_c = +5.05\,\xi_0$$

and, with a back substitution, the rotations:

$$\varphi_1 = +1.28\,\bar{\xi}_0$$
$$\varphi_2 = +2.92\,\bar{\xi}_0$$
$$\varphi_3 = +4.12\,\bar{\xi}_0$$

From this solution, one obtains, through the usual calculations that are here omitted, the diagrams of internal forces shown with continuous lines in Fig. 3.54.

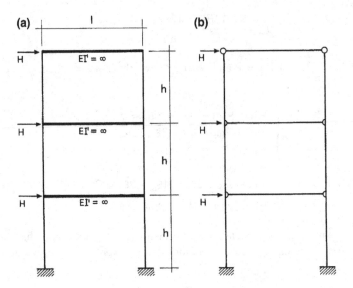

Fig. 3.53 Limit situations of **a** rigid beams and **b** hinged beams

It is interesting to compare the situation just examined, characterized by a flexural stiffness of the beam equal to that of the column, with the two limit situations described in Fig. 3.53a, b, the first one with rigid beams and the second with hinged beams.

In the situation with beams much more rigid than the columns ($k' = \infty$), the three rotation unknowns remain identically null and the solving algorithm is formed by only three equations that express the translation equilibrium of the beams:

$$\begin{cases} +\xi_a - \xi_b = \xi_o \\ -\xi_a + 2\xi_b - \xi_c = \xi_o \\ -\xi_b + 2\xi_c = \xi_o \end{cases}$$

where again it is set $\xi_o = Hh^3/24EI$. With a forward reduction, one obtains the solution:

$$\xi_c = 3\,\xi_o$$
$$\xi_b = 2\,\xi_o + \xi_c$$
$$\xi_a = 1\,\xi_o + \xi_b$$

that shows how every upper floor adds its translation deformation to that of the lower floors, deformation due to the pertinent horizontal force as received from the accumulation of the upper actions.

The bending moments on the columns are deduced from the deformation of their respective floor:

$$-M_{d3} = +M_{d1} = \frac{6EI}{h^2}1\xi_0 = \frac{1H}{2}\frac{h}{2} = V_d\frac{h}{2}$$

$$-M_{f5} = +M_{f3} = \frac{6EI}{h^2}2\xi_0 = \frac{2H}{2}\frac{h}{2} = V_f\frac{h}{2}$$

$$-M_{r7} = +M_{r5} = \frac{6EI}{h^2}3\xi_0 = \frac{3H}{2}\frac{h}{2} = V_r\frac{h}{2}$$

while the bending moments of the beams are obtained from the rotation equilibria of the joints:

$$M_{a1} = M_{d1} = +\frac{1}{4}Hh$$

$$M_{b3} = -M_{d3} + M_{f3} = +\frac{3}{4}Hh$$

$$M_{c5} = -M_{f5} + M_{r5} = +\frac{5}{4}Hh$$

The shear forces on the beams and the axial forces on all members are eventually calculated:

$$V_a = -\frac{M_{a1}}{1/2} = -2\frac{1}{4}H\frac{h}{1}$$

$$V_b = -\frac{M_{b3}}{1/2} = -2\frac{3}{4}H\frac{h}{1}$$

$$V_c = -\frac{M_{c5}}{1/2} = -2\frac{5}{4}H\frac{h}{1}$$

$$N_d = -V_a = +2\frac{1}{4}H\frac{H}{1}$$

$$N_f = N_d - V_b = +2\frac{4}{4}H\frac{H}{1}$$

$$N_r = N_f - V_c = +2\frac{9}{4}H\frac{H}{1}$$

$$N_a = N_b = N_c = -\frac{H}{2}$$

where the axial forces on the beams hold good on the assumption of horizontal actions applied wholly in the left joints 1, 2, 3 of the frame.

In the limit situation of hinged beams ($k' = 0$), these ones participate in the frame behaviour only ensuring an equal repartition of the horizontal forces on the two columns. So, each column behaves like an isostatic cantilever for which the internal forces can be immediately calculated with:

$$M'_d = -\frac{H}{2}h$$

$$M'_f = -\frac{H}{2}(1+2)h = -\frac{H}{2}3h$$

$$M'_r = -\frac{H}{2}(1+2+3)h = -\frac{H}{2}6h$$

$$V_d = 1\frac{H}{2}$$

$$V_f = 2\frac{H}{2}$$

$$V_r = 3\frac{H}{2}$$

$$N_d = N_f = N_r = 0$$

For what concerns the translations, with reference to the upper beam only, the calculation is carried out as for the isostatic cantilever of Sect. 2.1.3:

$$\xi_a = \frac{1}{2}\frac{Hh^3}{3EI}(3^3 + 2^3 + 1^3) + \frac{1}{2}\frac{Hh^2}{2EI}(2^2 \cdot 1 + 1^2 \cdot 2)h = \frac{15}{2}\frac{Hh^3}{EI} = \frac{180}{24}\frac{Hh^3}{EI}$$

Comparing this value with those of the preceding cases for which one had:

$$\xi_a = \frac{13.7}{24}\frac{Hh^3}{EI} \quad \text{for } k' = k$$

$$\xi_a = \frac{6}{24}\frac{Hh^3}{EI} \quad \text{for } k' = \infty$$

one has the following ratios:

$$\begin{aligned}
&1.00 \quad \text{for } k' = \infty \\
&2.28 \quad \text{for } k' = k \\
&30.00 \text{ for } k' = 0
\end{aligned}$$

From this comparison, one can notice the very big influence that the flexural stiffening given by the beams of the floors has on the translation deformability of the multi-storey frames.

In Fig. 3.54, the diagrams of internal forces for the two limit situations are drawn with dashed lines and marked, respectively, with $k' = \infty$ and $k = 0$. From this figure, one can notice how a flexural stiffness of the beams equal to the one of the columns is sufficient to obtain a big reduction of the bending moments of the columns with respect to those of the cantilever behaviour. This big reduction is paid with relevant axial forces on the columns and bending moments on the beams.

The most important contribution given to the columns by their full moment connection with the beams refers to the second-order effects including their possible

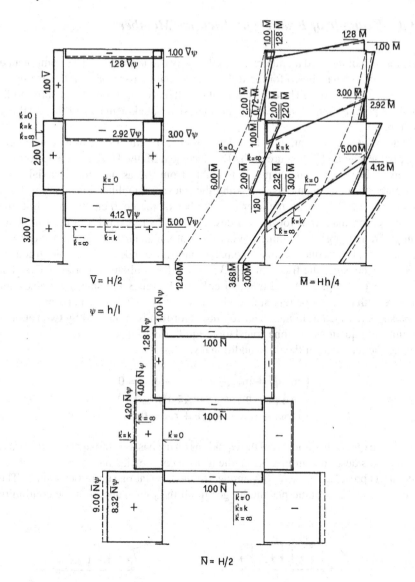

Fig. 3.54 Diagrams of internal forces

instability under the vertical "critical load". This aspect will be dealt with in deep in Chap. 4.

3.3.3 Example of Frame with Inclined Members

The presence of inclined members makes the sway frame analysis more complicated
with respect to what shown for the orthogonal member assemblies in the examples
of the preceding section. The calculation of the effects of the translations of the additional supports requires a more complex analysis of the kinematics of the auxiliary
structure.

In automated computation of however complex structures, it is more convenient
to apply the standard procedure of the Displacement Method (see Sect. 3.1.2), without any reduction of unknowns as derivable from the assumption of axial non-
deformability of the members, but using the relevant simplifications coming from
a standardized elaboration member by member of the coefficients, including the
possible co-ordinate transformations described in Sect. 3.1.3. From the following
example, remaining in the limited dimensions of a manual calculation, one can see
the operational consequences of a kinematic analysis extended to the whole structure.

So, let's consider the frame of Fig. 3.55a that has, unlike the portal analysed at
Sect. 2.3.3 (see Fig. 2.69), an inclined column. Figure 3.55b shows the auxiliary
structure with the rotation restraints added at joints 1 and 2 and the horizontal translation support added at joint 2. The solving algorithm is composed by two rotation
equilibrium equations and one translation equilibrium equation. They give a zero
value to the reactions of the three additional supports:

$$
\begin{cases}
m_{11}\varphi_1 + m_{12}\varphi_2 + m_{1x}\xi + m_{10} = 0 \\
m_{21}\varphi_1 + m_{22}\varphi_2 + m_{2x}\xi + m_{20} = 0 \\
r_{x1}\varphi_1 + r_{x2}\varphi_2 + r_{xx}\xi + r_{x0} = 0
\end{cases}
$$

Before expressing the terms of the equations on the basis of the usual superposition
of effects (as described in Fig. 3.57), the analysis of the kinematics of the auxiliary
structure is presented for a shift δ applied to the translation support of joint 2. The
structure is made isostatic placing hinges in all the joints, and on it the continuity

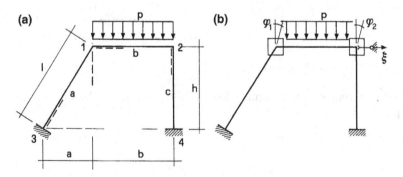

Fig. 3.55 Frame with an inclined member

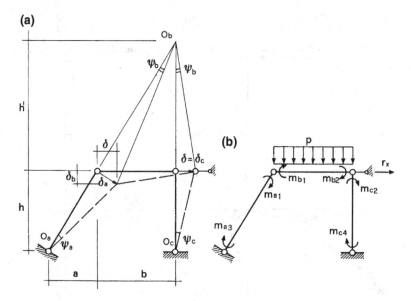

Fig. 3.56 Kinematics of the frame

moments are evidenced (positive when clockwise) as shown in Fig. 3.56b. For an applied shift δ, again with the assumption of null axial deformations of the members, a rigid displacement of the structure is displayed as shown in Fig. 3.56a. Around this configuration the flexural deformations of the members develop consistently to the rotation restraints present at their ends.

Within the field of small displacements, the rotations of the three members, around their centres O_a, O_b and O_c, are (with $h' = hb/a$):

$$\psi_c = \frac{\delta}{h}$$
$$\psi_b = \frac{\delta}{h'} = \frac{\delta}{h}\frac{a}{b}$$
$$\psi_a = \psi_c = \frac{\delta}{h}$$

and the orthogonal translations between the ends of the same members, written for $\delta = 1$, are:

$$\delta_c = 1$$
$$\delta_b = \psi_b\, b = a/h$$
$$\delta_a = \psi_a\, l = l/h$$

with $l = \sqrt{a^2 + h^2}$.

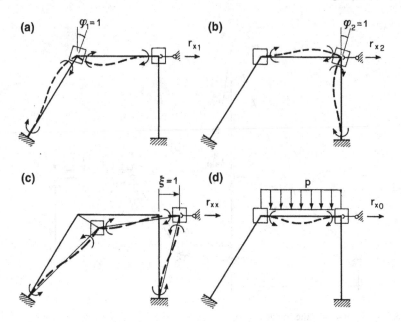

Fig. 3.57 Superposition of effects

After this introduction, the four cases described in Fig. 3.57 can be examined. The auxiliary structure has been submitted, one after the other, to the three support displacements corresponding to the geometric unity unknowns and to the applied load. Their effects are evidenced in the figure and listed hereunder with the assumption of members with constant cross section. The effects r_{xi} related to the reaction of the translation support are omitted: for these latter, there is a more convenient way that avoids the repeated projections of shear and axial forces.

– Rotation $\varphi_1 = 1$ (Fig. 3.57a):

$$m_{11} = +\frac{4EI_a}{1} + \frac{4EI_b}{b}$$
$$m_{21} = +\frac{2EI_b}{b}$$

– Rotation $\varphi_2 = 1$ (Fig. 3.57b):

$$m_{12} = +\frac{2EI_b}{b}$$
$$m_{22} = +\frac{4EI_b}{b} + \frac{4EI_c}{h}$$

– Translation $\xi = 1$ (Fig. 3.57c):

$$m_{1x} = -\frac{6EI_a}{l^2}\delta_a + \frac{6EI_b}{b^2}\delta_b = -\frac{6EI_a}{l^2}\frac{l}{h} + \frac{6EI_b}{b^2}\frac{a}{h}$$

$$m_{2x} = +\frac{6EI_b}{b^2}\delta_b - \frac{6EI_c}{h^2}\delta_c = +\frac{6EI_b}{b^2}\frac{a}{h} - \frac{6EI_c}{h^2}$$

– Applied load (Fig. 3.57d):

$$m_{10} = -\frac{pb^2}{12}$$

$$m_{20} = +\frac{pb^2}{12}$$

To express the reaction r_x of the additional translation support, let's consider the isostatic structure of Fig. 3.56b, where, following the same superposition of effects used above, the continuity moments are:

$$m_{a3} = +\frac{2EI_a}{l}\varphi_1 - \frac{6EI_a}{l^2}\frac{l}{h}\xi$$

$$m_{a1} = +\frac{4EI_a}{l}\varphi_1 - \frac{6EI_a}{l^2}\frac{l}{h}\xi$$

$$m_{b1} = +\frac{4EI_b}{b}\varphi_1 + \frac{2EI_b}{b}\varphi_2 + \frac{6EI_b}{b^2}\frac{a}{h}\xi - \frac{pb^2}{12}$$

$$m_{b2} = +\frac{2EI_b}{b}\varphi_1 + \frac{4EI_b}{b}\varphi_2 + \frac{6EI_b}{b^2}\frac{a}{h}\xi + \frac{pb^2}{12}$$

$$m_{c2} = +\frac{4EI_c}{h}\varphi_2 - \frac{6EI_c}{h^2}\xi$$

$$m_{c4} = +\frac{2EI_c}{h}\varphi_2 - \frac{6EI_c}{h^2}\xi$$

With the virtual work principle, an infinitesimal shift of the support is applied, obtaining the rigid deformation of the structure indicated in Fig. 3.56a. So, the following equation is set

$$r_x\delta + (m_{a3} + m_{a1})\psi_a - (m_{b1} + m_{b2})\psi_b + (m_{c2} + m_{c4})\psi_c + pb\frac{b}{2}\psi_b = 0$$

from which, with the pertinent substitutions and simplifications, one obtains for r_x the expression:

$$r_x = r_{x1}\varphi_1 + r_{x2}\varphi_2 + r_{xx}\xi + r_{x0}$$

where

$$r_{x1} = -\frac{6EI_a}{1}\frac{1}{h} + \frac{6EI_b}{b}\frac{a}{h}$$

$$r_{x2} = +\frac{6EI_b}{b^2}\frac{a}{h} - \frac{6EI_c}{h^2}$$

$$r_{xx} = +\frac{12EI_a}{1}\frac{1}{h^2} + \frac{12EI_b}{b^3}\frac{a^2}{h^2} + \frac{12EI_c}{h^3}$$

$$r_{x0} = -\frac{pb}{2}\frac{a}{h}$$

are the coefficients of the third equilibrium equation ($r_x = 0$).

Considering the equalities of the indirect coefficients:

$$r_{x1} = m_{1x}$$

$$r_{x2} = m_{2x}$$

the procedure could be reduced omitting, in the expressions of the moments m_{a1}, m_{a2}, ..., m_{c4}, the contributions of the rotations φ_1 and φ_2.

The solution is given for $I_a = I_b = I_c = I$ and for $b = h$, $a = 3\,h/4$, $l = 5\,h/4$. From the set of equations:

$$\begin{cases} 72\varphi_1 + 20\varphi_2 - 3\bar{\xi} = 10B \\ 4\varphi_1 + 16\varphi_2 - 3\bar{\xi} = -2B \\ -6\varphi_1 - 30\varphi_2 + 567\bar{\xi} = +90B \end{cases}$$

written with the pertinent simplifications and having set:

$$\bar{\xi} = \frac{\xi}{h}$$

$$B = \frac{pb^2}{12}\frac{h}{EI}$$

one obtains:

$$\varphi_1 = +0.1849\,B$$

$$\varphi_2 = -0.1425\,B$$

$$\xi = +0,1531\,Bh$$

From this solution, the moments are calculated ($\overline{M} = pb^2/12$):

$$M_{a3} = +m_{a3} = +\frac{8}{5}(0.1849 - 3 \cdot 0.1531)\overline{M} = -0.439\overline{M}$$

$$M_{a1} = -m_{a1} = -\frac{8}{5}(2 \cdot 0.1849 - 3 \cdot 0.1531)\overline{M} = +0.143\overline{M}$$

$$M_{b1} = -m_{b1} = -\frac{1}{4}(16 \cdot 0.1849 - 8 \cdot 0.1425 + 18 \cdot 0.1531 - 4)\overline{M} = +0.143\overline{M}$$

$$M_{b2} = -m_{b2} = -\frac{1}{4}(8 \cdot 0.1849 - 16 \cdot 0.1425 + 18 \cdot 0.1531 + 4)\overline{M} = -1.489\overline{M}$$

$$M_b = \frac{M_{b1} + M_{b2}}{2} + \frac{3}{2}\overline{M} = +0.827\overline{M}$$

$$M_{c2} = +m_{c2} = +2(-2 \cdot 0.1425 - 3 \cdot 0.1531)\overline{M} = -1.489\overline{M}$$

$$M_{c4} = -m_{c4} = -2(-0.1425 - 3 \cdot 0.1531)\overline{M} = +1.204\overline{M}$$

Figure 3.58 shows the correspondent diagrams. One can notice the fulfilled rotation equilibria at the joints 1 and 2 of the frame.

The shear forces (Fig. 3.58b) are calculated with the equilibria:

$$V_a = -\frac{m_{a3} + m_{a1}}{1} = +0.078\overline{V}$$

$$V'_b = +\overline{V} - \frac{m_{b1} + m_{b2}}{b} = +0.728\overline{V}$$

$$V''_b = -\overline{V} - \frac{m_{b1} + m_{b2}}{b} = -1.272\overline{V}$$

$$V_c = -\frac{m_{c2} + m_{c4}}{h} = +0.449\overline{V}$$

having set $\overline{V} = pb/2$.

The axial forces (Fig. 3.58c) are eventually calculated with:

$$N_c = +V''_b = -1.272\overline{N}$$

$$N_b = -V_c = -0.449\overline{N}$$

$$N_a = -V'_b\frac{h}{1} + N_b\frac{a}{1} = -0.852\overline{N}$$

having set $\overline{N} = pb/2$. With a projection of forces similar to the one written here above for N_a, one can verify the value of the shear force:

$$V_a = +V'_b\frac{a}{1} + N_b\frac{h}{1} = +0.078\overline{V}$$

that ensures the translation equilibrium of the frame.

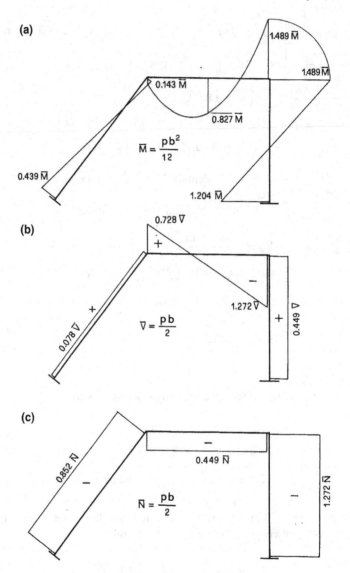

Fig. 3.58 Diagrams of the internal forces

3.4 Examples of Stiffness Analysis

In the structural analysis of the constructions, different possible load conditions are usually to be considered. The resistance verifications are then performed for any member under the heaviest conditions.

The search of the heaviest conditions, within a linear elastic behaviour, can rely usefully on the superposition of effects. The design can be in this way organized defining a number of *elementary load conditions* that refer to the different types of actions. For example, one can have:

- condition 1: self-weight of the structural elements (permanent);
- condition 2: (permanent) dead loads of the superimposed elements;
- condition 3: (variable) service live loads "a";
- condition 4: (variable) service live loads "b";
- condition i: ...
- condition n: wind pressure (variable).

To any elementary load condition corresponds a distribution of internal forces that is evaluated with a frame analysis similar to those presented in the preceding chapters. Eventually, in order to obtain the *critical situations* to be verified for resistance, the elementary conditions are to be "combined", taking into account the actual possibilities of coexistence of the loads and the different levels of probability related to their intensity.

The ordinary calculation procedure distinguishes different types of *load combinations* that are all based on proper summations of the related diagrams of internal forces, but differentiated for the weight given to the single contributions and for the orientation given to their accumulation. So in the following, the principal types of combinations are described. In the symbolic relations written below, S are the internal forces (bending moments, torsional moments, shear forces and axial forces). The summations are to be made "section by section".

The *simple combination* consists of the algebraic summation of two or more load conditions:

$$S_c = S_1 + S_2 + \cdots$$

and corresponds to the simple contemporaneity of more elementary load units.

In order to allow for the combination factors defined at Sect. 1.2.3 or because in the elementary conditions convenient fictitious values for the loads have been used (e.g. unity values), *weighted combinations* are computed:

$$S_c = \psi_1 S_1 + \psi_2 S_2 + \cdots$$

where any contribution has a weight ψ_i.

The *conditioned combination* is aimed at the accumulation to a pre-existent situation a new contribution only if it is pejorative, that is:

$$S_M = S_M + S \text{ if } S_M S > 0$$
$$S_M = S_M \quad\;\;\, \text{ if } S_M S < 0$$

just summing up section by section the effect of a variable action when it is unfavourable for resistance being of the same sign of the preceding one. The new contribution can be possibly weighted as in the weighted combination.

To be noted that the summations related to the combinations have to be extended to all the force components (e.g. *M, V, N*); the conditioning evaluation of the latter type of combination needs a *leading component* that drags the accumulation also of the other components: for example, an increased value of the bending moment could be calculated together with the corresponding shear and axial forces.

Also, the *selective combination* aims at summing up the contributions of the same sign, possibly weighted with the specific factor, but accumulating them in two different situations, the one S_{min} of the minimum force and the one S_{max} of the maximum force:

$$S_{min} = S_{min} + S \text{ if } S < 0$$
$$S_{max} = S_{max} + S \text{ if } S > 0$$

The selection evaluation is to be referred to a leading component as for the conditioned combination. But this latter gives the relative maximum value S_M, while the selective combination gives both the upper and the lower extremes of the variation field of the internal force.

For the progressive accumulations of the values of the conditioned and selective combinations, the response shall be "initialized"; that is, a *starting situation* shall be defined, in general corresponding to the effects of the permanent loads:

$$S_M = S_o$$

or

$$S_{min} = S_o$$
$$S_{max} = S_o$$

Eventually, the *substitutive combination* is quoted, usually referred to a pre-existing situation of positive or negative maximum force; it is aimed to replace the preceding value with the new one if the latter is heavier:

$$S_{min} = S \text{ if } S < S_{min}$$
$$S_{max} = S \text{ if } S > S_{max}$$

This combination refers to actions that cannot coexist, for example the "rightward wind" alternative to the "leftward wind".

Fig. 3.59 Example of portal structure

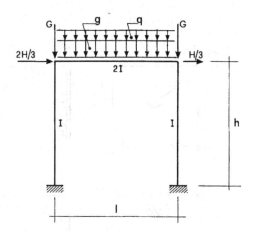

At the end of the combinations, one can obtain the *envelope diagrams* that delimit the family of critical situations deduced from the same combinations.

An example of action combination is shown with reference to the simple portal of Fig. 3.59, already analysed at Sect. 3.2.3 (Figs. 3.33 and 3.35 with $k_\varphi = \infty$). For the effects of the horizontal force, one can refer to Sect. 2.3.3 (Fig. 2.61 with $\kappa = 2$).

The following values are assumed:

$$l = h = 6 \text{ m}$$
$$g = 30 \text{ kN/m} \quad \text{permanent loads on the roof}$$
$$G = 30 \text{ kN} \quad \text{concentrated permanent loads}$$
$$q = 10 \text{ kN/m} \quad \text{variable loads on the roof}$$
$$H = 30 \text{ kN} \quad \text{wind pressure}$$

For these actions, Fig. 3.60 gives the three elementary conditions of internal forces:

$$S_1 \text{ "permanent"}$$
$$S_2 \text{ "variable"}$$
$$S_3 \text{ "wind"}$$

The prevalence of the permanent load effects could indicate that a reinforced concrete construction is dealt with. To be noted that the condition S_1 already represent a possible critical situation to be verified also if less heavy of the others. On the contrary, the load conditions S_2 and S_3 cannot exist by their selves.

The four possible load combinations are shown in Fig. 3.61:

$$S_4 = S_1 + S_2 \quad \text{"max. loads"}$$
$$S_5 = S_1 + S_3 \quad \text{"max. wind + min. load"}$$
$$S_6 = S_1 + \psi_{ov} S_3 \text{ "max. load + wind"}$$
$$S_7 = S_5 + \psi_{0q} S_2 \text{ "max. wind + loads"}$$

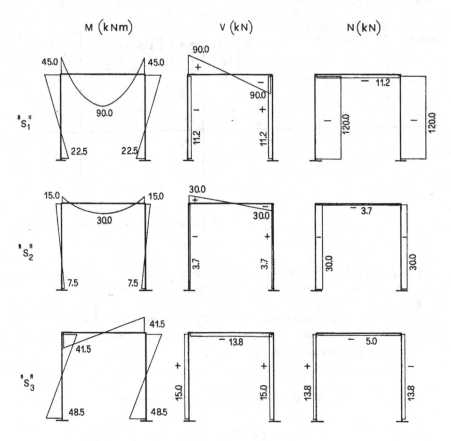

Fig. 3.60 Elementary conditions

The effects of the variable load and wind pressure are summed up to those of the permanent load in their different possible combinations. For the combination factors, the values $\psi_{ov} = 0.75$ and $\psi_{oq} = 0.60$ have been assumed. The combination leftwards wind has been omitted since it can be taken into account in the symmetric dimensioning of the portal.

The critical situations for resistance verification listed above have been obtained by simple and weighted combinations. In the following example that concludes the introduction of the present section, the envelope diagrams are instead elaborated through the proper sequence of weighted and selective combinations.

Reference is made to the continuous beam of Fig. 3.62 with:

$$l_a = 6.0 \text{ m}$$
$$l_b = 4.5 \text{ m}$$
$$g = 20 \text{ kN/m permanent load}$$
$$q = 20 \text{ kN/m variable load}$$

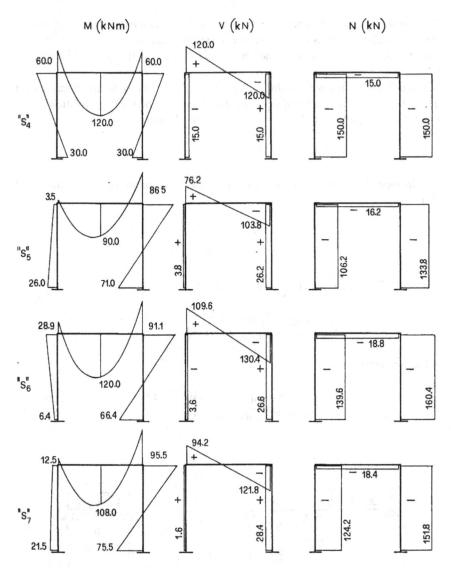

Fig. 3.61 Possible load combinations

Fig. 3.62 Example of two-span beam

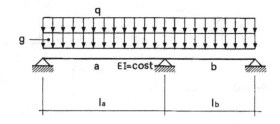

The lower relative relevance of the permanent load could indicate that a construction with steel structure is concerned. For the stress analysis, reference can be made to Sect. 2.4.2 (Fig. 2.98) and in particular to the "case 2" with $\kappa_0 = 0.75$ and $\kappa_0 = 0.33$.

The calculation starts from the elementary load conditions S_1 and S_2 as shown in Fig. 3.63 with unity value of the load. To obtain first the starting situation corresponding to the permanent loads, the weighted combination:

$$S_o = g\,S_1 + g\,S_2$$

is performed. Then, the situations of minimum and maximum stress are initialized with:

$$S_{min} = S_o$$
$$S_{max} = S_o$$

so to proceed with the selective accumulation of the contributions:

$$S_{min} = S_{min} + q\,S_1 \quad \text{if} \quad S_1 < 0$$
$$S_{max} = S_{max} + q\,S_1 \quad \text{if} \quad S_1 > 0$$

and subsequently:

$$S_{min} = S_{min} + q\,S_2 \quad \text{if} \quad S_2 < 0$$
$$S_{max} = S_{max} + q\,S_2 \quad \text{if} \quad S_2 > 0$$

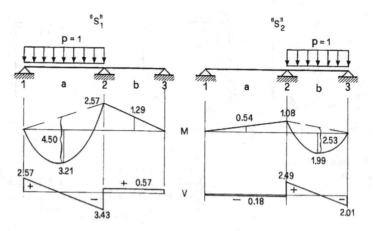

Fig. 3.63 Elementary load conditions

This accumulation is numerically carried out for the sections. 1, a, 2, b, 3 indicated in Fig. 3.63, with separate reference to the bending moment and the shear force, as if the related verifications could be disjointed.

	+64.2	−51.4	−25.8	$g\,S_1$
	−10.8	−21.6	+39.8	$g\,S_2$
	M_a=+53.4	M_2=−73.0	M_b=+14.0	S_o
+51.4	−68.6	+11.4	+11.4	$g\,S_1$
−3.6	−3.6	+49.8	−40.2	$g\,S_2$
V_1=47.8	V'_2=−72.2	V''_2=+61.2	V_3=−28.8	S_o
	+53.4	−73.0	+14.0	S_o
		−51.4	−25.8	$q\,S_1$
	−10.8	−21.6		$q\,S_2$
	M_a=+42.6	M_2=−146.0	M_b=−11.8	S_{min}
+47.8	−72.2	+61.2	−28.8	S_o
	−68.6			$q\,S_1$
−3.6	−3.6		−40.2	$q\,S_2$
V_1=+44.2	V'_2=−144.4	V''_2=+61.2	V_3=−69.0	S_{min}
	+53.4	−73.0	+14.0	S_o
	+64.2			$q\,S_1$
			+39.8	$q\,S_2$
	M_a=+117.6	M_2=−73.0	M_b=+53.8	S_{max}
+47.8	−72.2	+61.2	−28.8	S_o
+51.4		+11.4	+11.4	$q\,S_1$
		+49.8		$q\,S_2$
V_1=+99.2	V'_2=72.2	V''_2=122.4	V_3=−17.4	S_{max}

Figure 3.64 shows, other than the situation S_o of unloaded beam, the envelope diagrams drawn all along the beam. The lower S_{min} and upper S_{max} limits of the envelope are marked with thick lines, while thin lines indicate the completion parts of the diagrams. In the quoted figure, one can notice the set of the corresponding four critical situations for resistance verifications: *permanent loads, variable loads on span a, variable loads on span b* and *variable loads on spans a and b*.

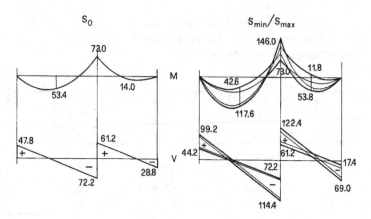

Fig. 3.64 Permanent and envelope diagram

3.4.1 Non-sway Frames

Reference is made to braced multi-storey building, for example those provided with a stiffening core such as the one described in Fig. 3.48b of Sect. 3.3.2. For the overall stability of these buildings against the horizontal actions, one can assume that these actions are wholly resisted by the stiffening core (see Figs. 3.49a and 3.50) and that, thanks to the superior stiffness of the core, the frames of beams and columns, constituting the bearing structures of the vertical loads, can be treated as non-sway frames.

Figure 3.65 shows an example of this type of multi-storey structure. The translation supports applied to the floors represent the horizontal bracing effect of the core. Actually, its effect is transmitted through the horizontal floor decks in a continuous way and not concentrated at one end as for simplicity sake is marked in the figure. Anyway, the position of these translation supports does not affect the flexural frame analysis presented here below.

Also if reduced to a plane model, the non-sway frame of Fig. 3.65, when globally analysed, leads to a large set of equations: following the Displacement Method, one has seven unknown rotations for each floor for a total amount of 35 unknowns.

Satisfactory approximate solutions can be obtained if, from the overall structural assembly, partial static schemes are extracted. Figure 3.65 indicates, for example, as a single floor beam could be examined independently from others, assuming proper interconnection supports. The type and the position of these supports are to be chosen so to provide a good approximation of the actual interconnection of the superimposed floors.

Figure 3.66 shows a partial frame of a floor, halved following the symmetry of the problem. This static scheme reproduces the full connections between beams and columns, and of these latter interprets in an approximate way the flexural behaviour with fixed hinges placed at mid-height.

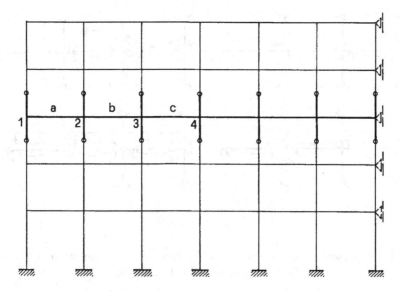

Fig. 3.65 Example of multi-storey frame structure

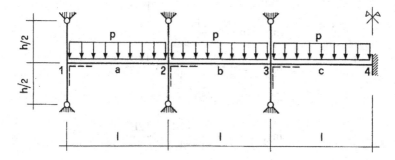

Fig. 3.66 Partial frame of a floor

For the solution, the additional rotation supports are applied at the joints 1, 2, 3 marking out the corresponding unknown rotations; the set of equilibrium equation is then expressed:

$$\begin{cases} m_{11}\varphi_1 + m_{12}\varphi_2 + m_{13}\varphi_3 + m_{10} = 0 \\ m_{21}\varphi_1 + m_{22}\varphi_2 + m_{23}\varphi_3 + m_{20} = 0 \\ m_{31}\varphi_1 + m_{32}\varphi_2 + m_{33}\varphi_3 + m_{30} = 0 \end{cases}$$

with the superposition of effects shown in Fig. 3.67.

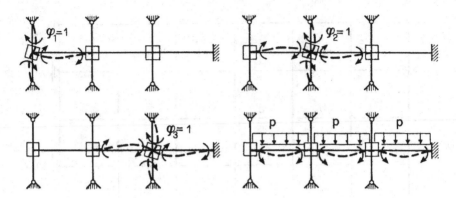

Fig. 3.67 Superposition of effects

Assuming a square mesh with $h = 1$ and for the beams a constant stiffness EI three time greater of the one EI' of the columns ($I' = I/3$), one has

$$m_{11} = 2\frac{3EI'}{h/2} + \frac{4EI}{1} = \frac{8EI}{1}$$

$$m_{12} = m_{21} = \frac{2EI}{1}$$

$$m_{22} = 2\frac{3EI'}{h/2} + 2\frac{4EI}{1} = \frac{12EI}{1}$$

$$m_{23} = m_{32} = \frac{2EI}{1}$$

$$m_{33} = 2\frac{3EI'}{h/2} + 2\frac{4EI}{1} = \frac{12EI}{1}$$

$$m_{13} = m_{31} = 0$$

$$m_{10} = -\frac{pl^2}{12} = -\frac{2}{3}\overline{M}$$

$$m_{20} = +\frac{pl^2}{12} - \frac{pl^2}{12} = 0$$

$$m_{30} = +\frac{pl^2}{12} - \frac{pl^2}{12} = 0$$

with

$$\overline{M} = \frac{pl^2}{8}$$

Through the opportune simplifications, the equations become

$$\begin{cases} 4\varphi_1 + \varphi_2 = \dfrac{1}{3EI}\overline{M} \\ \varphi_1 + 6\varphi_2 + \varphi_3 = 0 \\ \varphi_2 + 6\varphi_3 = 0 \end{cases}$$

and lead to the following solution:

$$\varphi_1 = +\frac{35}{134}\frac{1}{3EI}\overline{M}$$

$$\varphi_2 = -\frac{6}{134}\frac{1}{3EI}\overline{M}$$

$$\varphi_1 = +\frac{1}{134}\frac{1}{3EI}\overline{M}$$

From this solution, one obtains the moments:

$$M_{01} = -\frac{3EI'}{h/2}\varphi_1 = -\frac{35}{201}\overline{M}$$

$$M_{a1} = -\frac{2}{3}\overline{M} + \frac{4EI}{1}\varphi_1 + \frac{2EI}{1}\varphi_2 = -\frac{70}{201}\overline{M}$$

$$M_{a2} = -\frac{2}{3}\overline{M} - \frac{2EI}{1}\varphi_1 - \frac{4EI}{1}\varphi_2 = -\frac{157}{201}\overline{M}$$

$$M_{02} = -\frac{3EI'}{h/2}\varphi_2 = +\frac{6}{201}\overline{M}$$

$$M_{b2} = -\frac{2}{3}\overline{M} + \frac{4EI}{1}\varphi_2 + \frac{2EI}{1}\varphi_3 = -\frac{145}{201}\overline{M}$$

$$M_{b3} = -\frac{2}{3}\overline{M} - \frac{2EI}{1}\varphi_2 - \frac{4EI}{1}\varphi_3 = -\frac{130}{201}\overline{M}$$

$$M_{03} = -\frac{3EI'}{h/2}\varphi_3 = -\frac{1}{201}\overline{M}$$

$$M_{c3} = -\frac{2}{3}\overline{M} + \frac{4EI}{1}\varphi_3 = -\frac{132}{201}\overline{M}$$

$$M_{c4} = -\frac{2}{3}\overline{M} - \frac{2EI}{1}\varphi_3 = -\frac{135}{201}\overline{M}$$

$$M_a = \overline{M} + \frac{M_{a1} + M_{a2}}{2} = +\frac{175}{201}\frac{\overline{M}}{2}$$

$$M_b = \overline{M} + \frac{M_{b2} + M_{b3}}{2} = +\frac{127}{201}\frac{\overline{M}}{2}$$

$$M_c = \overline{M} + \frac{M_{c3} + M_{c4}}{2} = +\frac{135}{201}\frac{\overline{M}}{2}$$

where the index 0 indicates the moments on the columns. The shear forces are eventually calculated (with $\overline{V} = pl/2$):

$$V_{01} = +\frac{M_{01}}{h/2} = -\frac{70}{804}\overline{V}$$

$$V_{a1} = +\overline{V} - \frac{M_{a1} - M_{a2}}{l} = +\frac{717}{804}\overline{V}$$

$$V_{a2} = -\overline{V} - \frac{M_{a1} - M_{a2}}{l} = -\frac{891}{804}\overline{V}$$

$$V_{02} = +\frac{M_{02}}{h/2} = +\frac{12}{804}\overline{V}$$

$$V_{b2} = +\overline{V} - \frac{M_{b2} - M_{b3}}{l} = +\frac{819}{804}\overline{V}$$

$$V_{b3} = -\overline{V} - \frac{M_{b2} - M_{b3}}{l} = -\frac{789}{804}\overline{V}$$

$$V_{03} = +\frac{M_{03}}{h/2} = -\frac{2}{804}\overline{V}$$

$$V_{c3} = +\overline{V} - \frac{M_{c3} - M_{c4}}{l} = +\frac{801}{804}\overline{V}$$

$$V_{c4} = -\overline{V} - \frac{M_{c3} - M_{c4}}{l} = -\frac{807}{804}\overline{V}$$

Figure 3.68 shows the diagrams of the internal forces. One can notice first that on the internal columns very small flexural actions are transmitted due to the balanced actions of the two opposite beams. Only on the end column, a relevant flexural action is transmitted from the lateral unbalanced beam.

In Fig. 3.68, the diagrams obtained from the continuous beam model are also drawn with dashed lines. This model neglects the flexural contribution of the columns, interpreting the joint connections as simple supports (see Fig. 2.114 of 2.4.3). One can notice how for the internal beams the differences are very small, while for the end beam are more relevant. Anyhow, the continuous beam model leads, in the end beam, to higher bending moments, both positive at mid-span and negative at the internal support. The values of bending moment are inadequate approaching the end support where it becomes null.

From these considerations, it follows that the simplified scheme of continuous beam is acceptable, with its results presented at Sect. 2.4.3, provided a local correction is applied at the end parts. In the following, a procedure will be presented, again based on the analysis of a partial scheme, that allows to evaluate the bending moment at the end section of the beam and its distribution on the contiguous columns.

If, instead of the constant load distribution with the same value in all the beam spans, the partial load distributions are analysed, like those assumed in Figs. 2.115 and 2.116 of 2.4.3 with the aim of evaluating the maximum and minimum effects of the variable actions, more relevant differences of behaviour would be found between the schemes, with a higher engagement in the frame model of the flexural stiffness

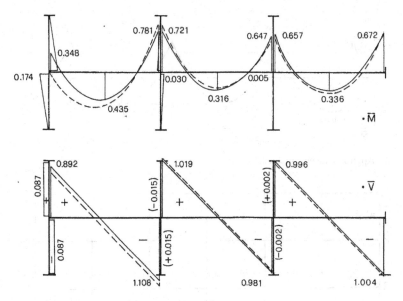

Fig. 3.68 Diagrams of bending moment and shear force

of the internal columns that would remain null in the continuous beam model. But in general, the conclusions presented above remain valid, with a good approximation of the continuous beam model in all the internal beam spans and columns subjected mainly to axial compression, and the necessity of a design improvement for the end beam spans and related end columns.

With reference only to the case of beam spans all equal here examined and to permanent loads value equal to that of the variable loads ($g = q$), calling $p = g + q$ the total load, the approximate equations deduced at Sect. 2.4.3 can be resumed (see Figs. 2.115 and 2.116), obtaining:

$$M_i = -\frac{1}{2}pl^2\left(\frac{1}{12.0} + \frac{1}{10.0}\right) \simeq -\frac{pl^2}{11}$$

$$M_0 = +\frac{1}{2}pl^2\left(\frac{1}{24.0} + \frac{1}{13.3}\right) \simeq +\frac{pl^2}{17}$$

$$M_i' = -\frac{1}{2}pl^2\left(\frac{1}{9.5} + \frac{1}{8,9}\right) \simeq -\frac{pl^2}{10}$$

$$M_0' = +\frac{1}{2}pl^2\left(\frac{1}{12.9} + \frac{1}{10.4}\right) \simeq +\frac{pl^2}{12}$$

These equations, with approximate rounded factors, can be usefully used in the proportioning design of the beams. They correspond to the envelope diagrams represented in Fig. 3.69. These diagrams have to be properly completed for the moments

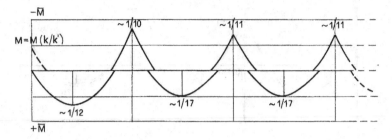

Fig. 3.69 Approximate envelope diagrams

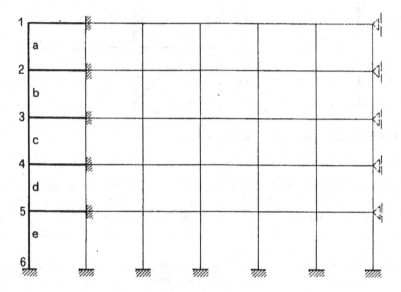

Fig. 3.70 Partial scheme for the end column

at the joints with the end columns, taking into account the actual ratio k/k' between the flexural stiffnesses of the elements.

Figure 3.70 shows how, from the multi-storey frame structure, a partial scheme can be extracted on which the behaviour of the end column and contiguous beams can be evaluated. The full supports at the internal end of these beams represent an extreme interpretation of their connection with the remaining part of the structure. This interpretation is not at the safe side. But the opposite extreme interpretation that would assume hinged connections, magnifying at the opposite end the continuity moments of the beams with the columns, seems excessively onerous with respect to the actual behaviour of the structure.

Figure 3.71a shows the partial scheme of concern. When assuming the results of the related analysis, it is to be reminded that the internal forces are approximated at the lower side. The equilibrium equations, with the usual procedure of the Displacement Method, are written as follows:

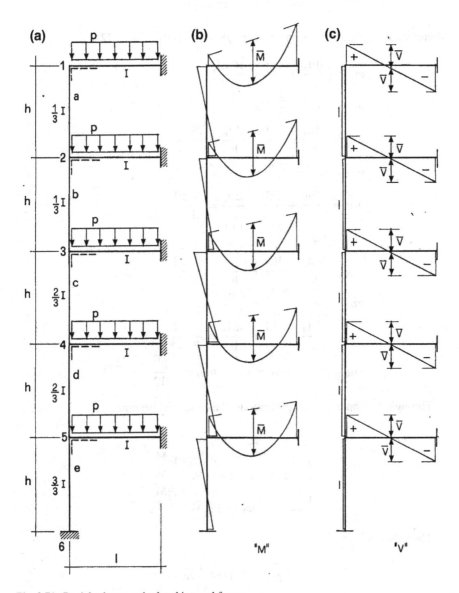

Fig. 3.71 Partial scheme and related internal forces

$$\begin{cases} m_{11}\varphi_1 + m_{12}\varphi_2 + m_{10} = 0 \\ m_{21}\varphi_1 + m_{22}\varphi_2 + m_{23}\varphi_3 + m_{20} = 0 \\ m_{32}\varphi_2 + m_{33}\varphi_3 + m_{34}\varphi_4 + m_{30} = 0 \\ m_{43}\varphi_3 + m_{44}\varphi_4 + m_{45}\varphi_5 + m_{40} = 0 \\ m_{54}\varphi_4 + m_{55}\varphi_5 + m_{50} = 0 \end{cases}$$

where, assuming again $h = 1$ and $\overline{M} = pl^2/8$, one has (see Fig. 3.72):

$$m_{11} = \frac{4EI_a}{h} + \frac{4EI}{1} = \frac{16}{3}\frac{EI}{1}$$

$$m_{12} = m_{21} = \frac{2EI_a}{h} = \frac{2}{3}\frac{EI}{1}$$

$$m_{22} = \frac{4EI_a}{h} + \frac{4EI}{1} + \frac{4EI_b}{h} = \frac{20}{3}\frac{EI}{1}$$

$$m_{23} = m_{32} = \frac{2EI_b}{h} = \frac{2}{3}\frac{EI}{1}$$

$$m_{33} = \frac{4EI_b}{h} + \frac{4EI}{1} + \frac{4EI_c}{h} = \frac{24}{3}\frac{EI}{1}$$

$$m_{34} = m_{43} = \frac{2EI_c}{h} = \frac{4}{3}\frac{EI}{1}$$

$$m_{44} = \frac{4EI_c}{h} + \frac{4EI}{1} + \frac{4EI_d}{h} = \frac{28}{3}\frac{EI}{1}$$

$$m_{45} = m_{54} = \frac{2EI_d}{h} = \frac{4}{3}\frac{EI}{1}$$

$$m_{55} = \frac{4EI_d}{h} + \frac{4EI}{1} + \frac{4EI_e}{h} = \frac{32}{3}\frac{EI}{1}$$

$$m_{10} = m_{20} = m_{30} = m_{40} = m_{50} = -\frac{pl^2}{12} = -\frac{2}{3}\overline{M}$$

Through the opportune simplifications, the equations become:

$$\begin{cases} 8\varphi_1 + 1\varphi_2 = +\frac{1}{EI}\overline{M} \\ 1\varphi_1 + 10\varphi_2 + 1\varphi_3 = +\frac{1}{EI}\overline{M} \\ 1\varphi_2 + 12\varphi_3 + 2\varphi_4 = +\frac{1}{EI}\overline{M} \\ 2\varphi_3 + 14\varphi_4 + 2\varphi_5 = +\frac{1}{EI}\overline{M} \\ 2\varphi_4 + 16\varphi_5 = +\frac{1}{EI}\overline{M} \end{cases}$$

The solution is:

$$\varphi_1 = +0.1184\frac{1}{EI}\overline{M}$$

$$\varphi_2 = +0.0818\frac{1}{EI}\overline{M}$$

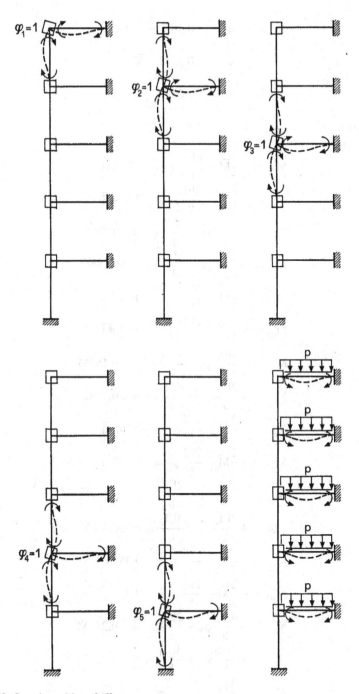

Fig. 3.72 Superimposition of effects

$$\varphi_3 = +0.0676 \frac{1}{EI} \overline{M}$$

$$\varphi_4 = +0.0538 \frac{1}{EI} \overline{M}$$

$$\varphi_5 = +0.0558 \frac{1}{EI} \overline{M}$$

and leads to the moments:

$$M_{01} = -\frac{2}{3}\overline{M} + \frac{4EI}{l}\varphi_1 = -0.208\overline{M}$$

$$M'_{01} = -\frac{2}{3}\overline{M} - \frac{2EI}{l}\varphi_1 = -0.896\overline{M}$$

$$M_{a1} = -\frac{4EI_a}{h}\varphi_1 - \frac{2EI_a}{h}\varphi_2 = -0.208\overline{M}$$

$$M_{a2} = +\frac{2EI_a}{h}\varphi_1 + \frac{4EI_a}{h}\varphi_2 = +0.186\overline{M}$$

$$M_{02} = -\frac{2}{3}\overline{M} + \frac{4EI}{l}\varphi_2 = -0.340\overline{M}$$

$$M'_{02} = -\frac{2}{3}\overline{M} - \frac{2EI}{l}\varphi_2 = -0.830\overline{M}$$

$$M_{b2} = -\frac{4EI_b}{h}\varphi_2 - \frac{2EI_b}{h}\varphi_3 = -0.154\overline{M}$$

$$M_{b3} = +\frac{2EI_b}{h}\varphi_2 + \frac{4EI_b}{h}\varphi_3 = +0.145\overline{M}$$

$$M_{03} = -\frac{2}{3}\overline{M} + \frac{4EI}{l}\varphi_3 = -0.396\overline{M}$$

$$M'_{03} = -\frac{2}{3}\overline{M} - \frac{2EI}{l}\varphi_3 = -0.802\overline{M}$$

$$M_{c3} = -\frac{4EI_c}{h}\varphi_3 - \frac{2EI_c}{h}\varphi_4 = -0.252\overline{M}$$

$$M_{c4} = +\frac{2EI_c}{h}\varphi_3 + \frac{4EI_c}{h}\varphi_3 = +0.234\overline{M}$$

$$M_{04} = -\frac{2}{3}\overline{M} + \frac{4EI}{l}\varphi_4 = -0.451\overline{M}$$

$$M'_{04} = -\frac{2}{3}\overline{M} - \frac{2EI}{l}\varphi_4 = -0.774\overline{M}$$

$$M_{d4} = -\frac{4EI_d}{h}\varphi_4 - \frac{2EI_d}{h}\varphi_5 = -0.218\overline{M}$$

$$M_{d5} = +\frac{2EI_d}{h}\varphi_4 + \frac{4EI_d}{h}\varphi_5 = +0.220\overline{M}$$

$$M_{05} = -\frac{2}{3}\overline{M} + \frac{4EI}{l}\varphi_5 = -0.444\overline{M}$$

$$M'_{05} = -\frac{2}{3}\overline{M} - \frac{2EI}{l}\varphi_5 = -0.778\overline{M}$$

$$M_{e5} = -\frac{4EI_e}{h}\varphi_5 = -0.223\overline{M}$$

$$M_{e6} = +\frac{2EI_e}{h}\varphi_5 = +0.112\overline{M}$$

In the equations above, M_{0i} and M'_{0i} indicate the moments at the left and right ends of the beams. The shear forces are calculated below (with $\overline{V} = pl/2$):

$$V_{01} = +\overline{V} - \frac{M_{01} - M'_{01}}{l} = +0.828\overline{V}$$

$$V_a = -\frac{M_{a2} - M_{a1}}{h} = -0.099\overline{V}$$

$$V_{02} = +\overline{V} - \frac{M_{02} - M'_{02}}{l} = +0.878\overline{V}$$

$$V_b = -\frac{M_{b3} - M_{b2}}{h} = -0.075\overline{V}$$

$$V_{03} = +\overline{V} - \frac{M_{03} - M'_{03}}{l} = +0.899\overline{V}$$

$$V_c = -\frac{M_{c4} - M_{c3}}{h} = -0.122\overline{V}$$

$$V_{04} = +\overline{V} - \frac{M_{04} - M'_{04}}{l} = +0.919\overline{V}$$

$$V_d = -\frac{M_{d5} - M_{d4}}{h} = -0.110\overline{V}$$

$$V_{05} = +\overline{V} - \frac{M_{05} - M'_{05}}{l} = +0.917\overline{V}$$

$$V_e = -\frac{M_{e6} - M_{e5}}{h} = -0.084\overline{V}$$

where V_{0i} indicates the shear force at the left end of the beams.

The whole diagrams of the internal forces calculated above are shown in Fig. 3.71b, c. Obviously for the verifications of the columns, the fundamental axial component N shall be added. For the beams in general, this component that comes from the difference between the shear forces of the upper and lower columns is little and does not affect sensibly their flexural verifications.

3.4.2 Sway Frames

In what follows, two examples of sway frames analysed with the Displacement Method are presented, remaining again within the approximation that neglects the axial deformations of the members.

Asymmetric Portal

For the solution of the portal of Fig. 3.73a, the auxiliary structure of Fig. 3.73b is assumed, where the geometric unknowns φ_1, φ_2 and ξ are indicated. Figure 3.74 shows the superposition of effects that leads to the following equilibrium equations:

$$\begin{cases} m_{11}\varphi_1 + m_{12}\varphi_2 + m_{1x}\xi + m_{10} = 0 \\ m_{21}\varphi_1 + m_{22}\varphi_2 + m_{2x}\xi + m_{20} = 0 \\ r_{x1}\varphi_1 + r_{x2}\varphi_2 + r_{xx}\xi + r_{x0} = 0 \end{cases}$$

The coefficients of the equations are given by the following expressions:

$$m_{11} = \frac{4EI}{l} + \frac{4EI'}{h}$$

$$m_{12} = m_{21} = \frac{2EI}{l}$$

$$m_{1x} = r_{x1} = -\frac{6EI'}{h^2}$$

$$m_{22} = \frac{4EI}{l} + \frac{4EI'}{s}$$

$$m_{2x} = r_{x2} = -\frac{6EI'}{s^2}$$

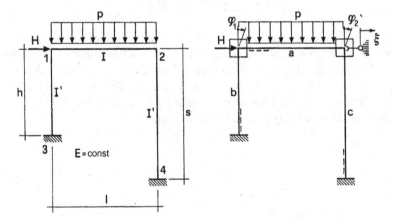

Fig. 3.73 Asymmetric portal

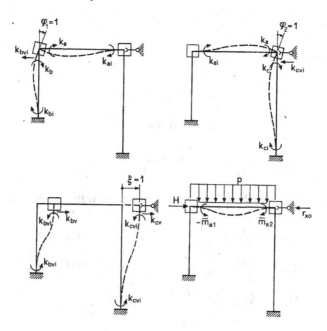

Fig. 3.74 Superposition of effects

$$r_{xx} = \frac{12EI'}{h^3} + \frac{12EI'}{s^3}$$

Assuming $h = 4\,1/5$, $s = 6\,1/5$ and $I' = I/2$, after the opportune simplifications the equations become (with $\bar{\xi} = \xi/l$):

$$\begin{cases} 104\varphi_1 + 32\varphi_2 - 75\bar{\xi} = -16\dfrac{1}{EI}m_{10} \\[3mm] 72\varphi_1 + 204\varphi_2 - 75\bar{\xi} = -36\dfrac{1}{EI}m_{20} \\[3mm] -1350\varphi_1 - 600\varphi_2 + 4375\bar{\xi} = -288\dfrac{l^2}{EI}r_{x0} \end{cases}$$

For the "vertical load" condition, with $p \neq 0$ and $H = 0$, setting $\overline{M} = pl^2/8$ one has:

$$m_{10} = -\frac{pl^2}{12} = -\frac{2}{3}\overline{M}$$

$$m_{20} = +\frac{pl^2}{12} = +\frac{2}{3}\overline{M}$$

$$r_{x0} = 0$$

and the solution is:

$$\varphi_1 = +0.1772 \frac{1}{EI}\overline{M}$$

$$\varphi_2 = -0.1686 \frac{1}{EI}\overline{M}$$

$$\overline{\xi} = +0.0316 \frac{1}{EI}\overline{M}$$

This solution leads to the following internal forces (with $\overline{V} = \overline{N} = pl/2$):

$$M_{a1} = -\frac{2}{3}\overline{M} + \frac{4EI}{l}\varphi_1 + \frac{2EI}{l}\varphi_2 = -0.295\overline{M} = M_{b1}$$

$$M_{a2} = -\frac{2}{3}\overline{M} - \frac{2EI}{l}\varphi_1 - \frac{4EI}{l}\varphi_2 = -0.347\overline{M} = M_{c2}$$

$$M_a = +\overline{M} + \frac{M_{a1} + M_{a2}}{2} = +0.679\overline{M}$$

$$M_{b3} = +\frac{2EI'}{h}\varphi_1 + \frac{6EI'}{h^2}\xi = +0.073\overline{M}$$

$$M_{c4} = -\frac{2EI'}{s}\varphi_2 + \frac{6EI'}{s^2}\xi = +0.206\overline{M}$$

$$V'_a = +\overline{V} - \frac{M_{a1} - M_{a2}}{l} = +0.987\overline{V} = -N_b$$

$$V''_a = -\overline{V} - \frac{M_{a1} - M_{a2}}{l} = +1.013\overline{V} = +N_c$$

$$V_b = -\frac{M_{b3} - M_{b1}}{h} = -0.115\overline{V} = +N_a$$

$$V_c = -\frac{M_{c2} - M_{c4}}{s} = +0.115\overline{V} = -N_a$$

Figure 3.75 shows, other than the deformed shape of the structure, the diagrams of the internal forces calculated above. One can notice that the structural dissymmetry leads, under a symmetric vertical load, to a translation of the beam oriented in the direction of the action of the column with higher flexural stiffness.

For the "horizontal thrust" condition, with $p = 0$ and $H \neq 0$, one has:

$$m_{10} = m_{20} = 0$$

$$r_{x0} = -H$$

and from the equations, with $\overline{M} = Hl$, one has the solution:

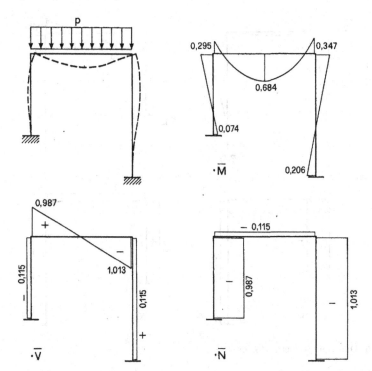

Fig. 3.75 Internal forces for vertical load

$$\varphi_1 = +0.05776\,\frac{1}{EI}\,\overline{M}$$

$$\varphi_2 = +0.01074\,\frac{1}{EI}\,\overline{M}$$

$$\overline{\xi} = +0.08467\,\frac{1}{EI}\,\overline{M}$$

that leads to the internal forces:

$$M_{a1} = +\frac{4EI}{l}\varphi_1 + \frac{2EI}{l}\varphi_2 = +0.253\overline{M} = M_{b1}$$

$$M_{a2} = -\frac{2EI}{l}\varphi_1 - \frac{4EI}{l}\varphi_2 = -0.158\overline{M} = M_{c2}$$

$$M_{b3} = +\frac{2EI'}{h}\varphi_1 - \frac{6EI'}{h^2}\xi = -0.325\overline{M}$$

$$M_{c4} = -\frac{2EI'}{s}\varphi_2 + \frac{6EI'}{s^2}\xi = +0.167\overline{M}$$

$$V_a = -\frac{M_{a1} - M_{a2}}{l} = -0.411H = -N_b = +N_c$$

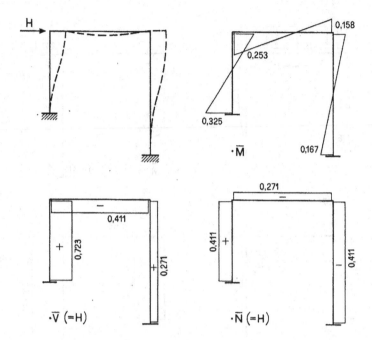

Fig. 3.76 Diagrams of internal forces

$$V_b = -\frac{M_{b3} - M_{b1}}{h} = +0.723H$$

$$V_c = -\frac{M_{c2} - M_{c4}}{s} = +0.271H = -N_a$$

Figure 3.76 shows, other than the deformed shape of the structure, the diagrams of the internal forces calculated above. One can notice that the horizontal action is mainly resisted by the stiffer column.

Portal with Inclined Pitches

For the analysis of the portal of Fig. 3.77 following the standard procedure of the Displacement Method, five geometric unknowns would be needed: the three rotations of the internal joints, plus the two translations necessary to render determined the kinematics of the frame (e.g. the horizontal translations of the joints 1 and 3). For a manual calculation, it would be convenient to apply another method that, like the one based on the virtual work principle, requires a number of unknowns equal to the redundancies of the given structure, for example the three components of the reaction of an external support.

In these pages, the analysis is performed only under two load conditions, one symmetric and one anti-symmetric, for which the problem is sensibly reduced.

For the "vertical load" condition, one can exploit the symmetry of the structural arrangement for which only two geometrical unknowns remain:

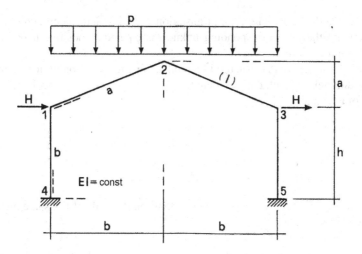

Fig. 3.77 Portal with inclined pitches

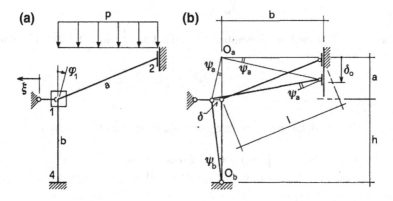

Fig. 3.78 Auxiliary structure

$$\varphi_1 = -\varphi_3$$
$$(\varphi_2 = 0)$$
$$\xi_1 = -\xi_3$$

So, one can refer to the halved frame of Fig. 3.78a, with its support 2 consistent with the symmetry conditions:

$$\begin{array}{ll} \varphi_2 = 0 & m_2 \neq 0 \\ \xi_2 = 0 & r_{x2} \neq 0 \\ \eta_2 \neq 0 & r_{y2} = 0 \end{array}$$

The two additional rotation and translation restraints have been added to joint 1, evidencing the two corresponding unknowns φ_1 and ξ, positive if oriented as indicated in the figure.

Figure 3.78b shows the structural kinematics due to a horizontal displacement δ of the additional restraint: the two members "a" and "b" rotate around the respective centres of an angle:

$$\psi_a = \frac{\delta}{a}$$

$$\psi_b = \frac{\delta}{h}$$

and the orthogonal relative displacements between the ends of the same members, evaluated for $\delta = 1$, are:

$$\delta_a = \psi_a 1 = \frac{1}{a}$$

$$\delta_b = 1$$

The lowering of the loaded member "a" varies linearly from 0 at its left end to:

$$\delta_0 = \psi_a b = \frac{b}{a}\delta$$

at its right end.

This being stated, the equilibrium equations can be written as follows:

$$\begin{cases} m_{11}\varphi_1 + m_{1x}\xi + m_{10} = 0 \\ r_{x1}\varphi_1 + r_{xx}\xi + r_{x0} = 0 \end{cases}$$

For the evaluation of the coefficients, reference is made to the superposition of effects represented in Fig. 3.79a–c, from which, for a rotation $\varphi_1 = 1$, one obtains the moments:

$$m_{11} = +\frac{4EI}{h} + \frac{4EI}{1}$$

$$m_{21} = +\frac{2EI}{1}$$

$$m_{41} = +\frac{2EI}{h}$$

for a translation $\xi = 1$, one obtains the moments:

$$m_{1x} = +\frac{6EI}{h^2}\delta_b - \frac{6EI}{1^2}\delta_a = +\frac{6EI}{h^2} - \frac{6EI}{al} = r_{x1}$$

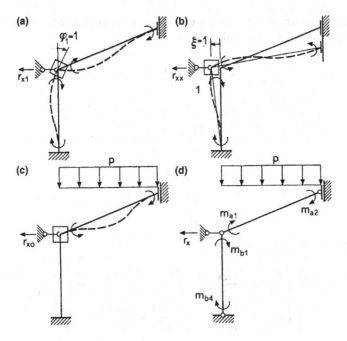

Fig. 3.79 Superimposition of effects

$$m_{2x} = -\frac{6EI}{l^2}\delta_a = -\frac{6EI}{al}$$
$$m_{4x} = +\frac{6EI}{h^2}\delta_b = +\frac{6EI}{h^2}$$

and for the action of the external load, one obtains the moments:

$$m_{10} = -\frac{pb^2}{12}$$
$$m_{20} = +\frac{pb^2}{12}$$
$$m_{40} = 0$$

Placing now these moments on the isostatic structure of Fig. 3.79d to represent the flexural continuity at the joint, for the calculation of the reaction r_x the virtual work principle is used applying an infinitesimal displacement δ to the support (see Fig. 3.78b) and the work of the applied forces is made equal to zero:

$$r_x\delta + (m_{a1} + m_{a2})\psi_a - (m_{b1} + m_{b4})\psi_b + pb\frac{b}{2}\psi_a = 0$$

From this equation, simplifying the quantity δ and after the pertinent substitutions and simplifications, one eventually obtains:

$$r_x = r_{x1}\,\varphi_1 + r_{xx}\,\xi + r_{x0}$$

where

$$r_{x1} = +\frac{6EI}{h^2} - \frac{6EI}{al} = m_{1x}$$

$$r_{xx} = +\frac{12EI}{h^3} + \frac{12EI}{a^2 l}$$

$$r_{x0} = -\frac{pb^2}{2a}$$

So, one has the two equations:

$$\begin{cases} \left(\dfrac{4EI}{h} + \dfrac{4EI}{l}\right)\varphi_1 + \left(\dfrac{6EI}{h^2} - \dfrac{6EI}{al}\right)\xi = \dfrac{pb^2}{12} \\[2mm] \left(\dfrac{6EI}{h^2} + \dfrac{6EI}{al}\right)\varphi_1 + \left(\dfrac{12EI}{h^3} + \dfrac{12EI}{a^2 l}\right)\xi = \dfrac{pb^2}{2a} \end{cases}$$

The solution is elaborated for $h = b$, $a = 0.458b$ and $l = 1.100b$. Setting $\overline{M} = pb^2/12$, after the opportune simplifications the equations become (with $\bar{\xi} = \xi/h$):

$$\begin{cases} +3.818\varphi_1 - 2.951\bar{\xi} = \dfrac{h}{2EI}\overline{M} \\[2mm] -2.951\varphi_1 + 31.974\bar{\xi} = 13.093\dfrac{h}{2EI}\overline{M} \end{cases}$$

and the solution is:

$$\varphi_1 = +0.6229\frac{h}{2EI}\overline{M}$$

$$\xi = +0.4670\frac{h^2}{2EI}\overline{M}$$

The internal forces are eventually calculated as follows (with $\overline{V} = \overline{N} = pb$):

$$M_{a2} = -\overline{M} - \frac{2EI}{l}\varphi_1 + \frac{6EI}{al}\xi = +1.213\overline{M}$$

$$M_{b1} = -\frac{4EI}{h}\varphi_1 - \frac{6EI}{h^2}\xi = -2.647\overline{M} \quad (= M_{a1})$$

$$M_{b4} = +\frac{2EI}{h}\varphi_1 + \frac{6EI}{h^2}\xi = +2.024\overline{M}$$

$$N_b = -pb = -\overline{N}$$

$$V_b = -\frac{M_{b4} - M_{b1}}{h} = -0.389\overline{V}$$

$$V'_a = -N_b \frac{b}{l} + V_b \frac{a}{l} = +0.747\overline{V}$$

$$V''_a = +V'_a - pb\frac{b}{l} = -0.162\overline{V}$$

$$N'_a = +N_b \frac{a}{l} + V_b \frac{b}{l} = -0.770\overline{N}$$

$$N''_a = N'_a + pb\frac{a}{l} = -0.354\overline{N}$$

The corresponding diagrams, completed over all the structures, are shown in Fig. 3.80. In particular, the bending moment at mid-span of the pitches is obtained with:

$$M_a = \frac{3}{2}\overline{M} + \frac{M_{a1} + M_{a2}}{2} = +0.783\overline{M}$$

For the "horizontal thrust" condition, the anti-symmetry of the structural arrangement is exploited from which only two unknowns remain:

$$\varphi_1 = \varphi_3$$
$$\varphi_2 = -\varphi_1/2$$
$$\xi_1 = \xi_3 = \xi$$

So, one can refer to the halved frame of Fig. 3.81a, with the support at joint 2 consistent with the conditions of anti-symmetry:

$$\varphi_2 \neq 0 \qquad m_2 = 0$$
$$\xi_2 \neq 0 \qquad r_{x2} = 0$$
$$\eta_2 = 0 \qquad r_{y2} \neq 0$$

The two rotation and translation restraints at joint 1 have been added evidencing the correspondent unknown φ_1 and ξ, positive if oriented as indicated in the figure.

Figure 3.81b shows the structural kinematics due to a horizontal displacement δ of the additional restraint. The member "b" rotates by an angle:

$$\psi_b = \frac{\delta}{h}$$

around its centre O_b, while the member "a" displays a simple translation ($\psi_a = 0$). This being stated, the equilibrium equations can be written as follows:

$$\begin{cases} m_{11}\varphi_1 + m_{1x}\xi + m_{10} = 0 \\ r_{x1}\varphi_1 + r_{xx}\xi + r_{x0} = 0 \end{cases}$$

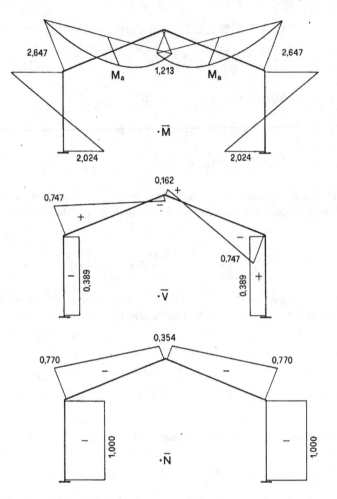

Fig. 3.80 Diagrams of internal forces

where the coefficients are (see Fig. 3.82):

$$m_{11} = +\frac{4EI}{h} + \frac{3EI}{l}$$

$$r_{x1} = -\frac{6EI}{h^2} = m_{1x}$$

$$r_{xx} = +\frac{12EI}{h^3}$$

$$m_{10} = 0$$

$$r_{x0} = -H$$

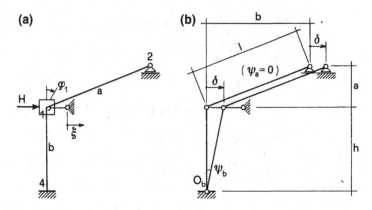

Fig. 3.81 Auxiliary structure

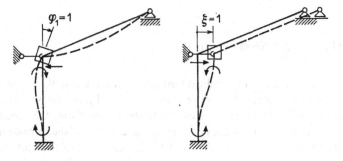

Fig. 3.82 Superimposition of effects

With the same data assumed in the preceding case ($h = b$, $a = 0.458b$ and $1 = 1.100b$), one obtains the equations

$$\begin{cases} +1.121\varphi_1 - 1.000\bar{\xi} = 0 \\ -1.000\varphi_1 + 2.000\bar{\xi} = \dfrac{h}{6EI}\overline{M} \end{cases}$$

with $\overline{M} = Hh$ and $\bar{\xi} = \xi/h$. The solution is:

$$\varphi_1 = +0.1341\frac{h}{EI}\overline{M}$$

$$\xi = +0.1504\frac{h^2}{EI}\overline{M}$$

The internal forces are eventually calculated:

$$M_{a1} = +\frac{3EI}{1}\varphi_1 = +0.366\overline{M}$$

$$M_{b1} = -\frac{4EI}{h}\varphi_1 + \frac{6EI}{h^2}\xi = +0.366\overline{M}$$

$$M_{b4} = +\frac{2EI}{h}\varphi_1 - \frac{6EI}{h^2}\xi = -0.634\overline{M}$$

$$V_b = +H$$

$$V_a = -\frac{M_{a1}}{1} = -0.333H$$

$$N_a = -V_a\frac{a}{b} = +0.152H$$

$$N_b = +\frac{M_{a1}}{b} = +0.366H$$

Figure 3.83 shows the diagrams of the internal forces where in particular $\overline{V} = \overline{N} = H$ has been set.

3.4.3 Qualitative Stress Diagrams

After the solution of a frame, obtained through a manual elaboration following the methods presented in these chapters or using automated procedures of calculation, opportune checks on the correctness of results shall be performed. So assuming to have available, other than the values of joint displacements and the force components on the end sections of the members, also the diagrams of the internal forces over all the structures, one has to verify the consistency of these results with the input data of the given problem as specified in the structural static scheme.

First, it is possible to *verify the equilibrium of joints* and this ensures the correct solution of the equations and the correct insertion of the involved data. One can also *verify the global equilibrium of members*, with vectorial summations of the applied loads with the reactions rendered to the same members through their end sections. With reference to the reactions of the external supports, one can also *verify the global equilibrium of the structure*. Eventually, one can *verify the local equilibrium of the current sections*. This last check is ruled by the elastic line relations and, for example, leads to parabolic bending moment diagrams on beam segments subjected to distributed flexural load of constant intensity and leads to angular discontinuities in the same diagram under concentrated loads and linear variations along unloaded segments. With reference to the relation between the shear force and the variation of the bending moment, the linear, discontinuous or constant shape of the diagram of shear force shall be verified under the same situations quoted above or null along the beam segments with constant bending moment.

The check of geometric compatibility is in general more difficult. Usually, the deformed shape of the structure, drawn throughout all the structural assemblies, is not available. One has only the numerical values of the joint displacements. From a structural analysis made with the Displacement Method, these values come out

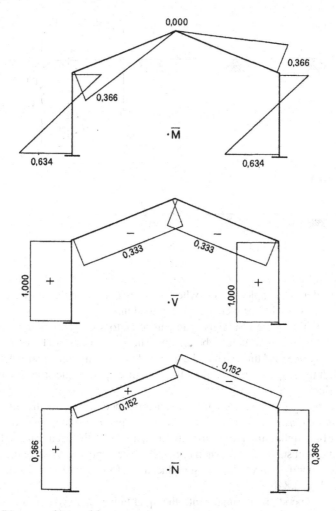

Fig. 3.83 Diagrams of internal forces

implicitly consistent with the local joint geometric compatibility. One could check this compatibility at the external supports if the results come from an automated calculation procedure and correctness of the topological input data is to be verified. For some simple situations, where there are not relevant interactions among different loads, it is possible to *verify the direct effects* that should lead to displacements oriented in the same direction of the applied loads. One can also *verify the symmetry and anti-symmetry conditions* possibly present in the structure and *verify the similarity with elementary well-known situations.*

For the correct use of the results of the structural analysis, the experience formed with the *drawing of qualitative diagrams of internal forces* is very useful. This is

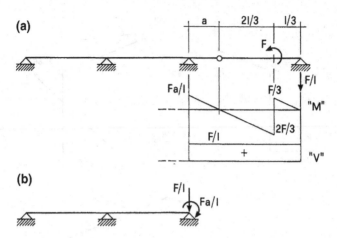

Fig. 3.84 Continuous beam with an isostatic part

easy if one refers to simple frames, where there is not the problem of the possible interactions of many different contemporary load units.

In the drawing of the diagrams, one can refer to some *typical situations* that contribute to the overall structural behaviour. These situations are here presented in the case of only one load unit, so that the origin of the action is univocally determined together with the response rendered by the remaining part of the structure to ensure the equilibrium.

The first consideration refers to the possible presence, within the structural assembly, of *isostatic parts*. These parts can be treated separately, independently from the behaviour of the remaining part of the structure, applying the usual cardinal equilibrium conditions of statics, as shown in Fig. 3.84a. The hyperstatic part of the structure can be subsequently analysed, having replaced the former part with its determined actions (see Fig. 3.84b).

Within a hyperstatic arrangement, the qualitative analysis of the structural behaviour starts from its *loaded member*. As shown in Fig. 3.85a, for this member it is possible to place the response within the *limit situations* correspondent to extreme interpretations of the end supports. If the remaining part of the structure is considered very stiff, the diagram of the bending moment is that marked with (1) related to a full end support. On the contrary, if the remaining part of the structure is considered very flexible, the diagram is that marked with (2) related to a simple end support. The actual response, as indicated in Fig. 3.85a, lays in an intermediate position, closer to one or the other limit depending on the stiffness ratio between the parts.

Evaluated in this way the response of the loaded member, one has subsequently to analyse how the actions propagate on the remaining part of the structure. For a serial sequence of members, like the one of Fig. 3.85, the *stress transmission* displays

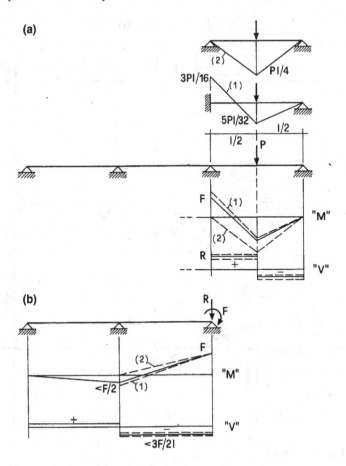

Fig. 3.85 Hyperstatic system with one loaded member

a progressive damping with the distance that can be estimated with the following procedure.

The loaded member is replaced by the action transmitted to the remaining part of the structure: for example, with the moment F (<3Pl/16) of Fig. 3.85b. The response of the adjacent member can be placed between the limit situations (1) and (2) corresponding, respectively, to an opposite full or simple end support. Within these two limit situations lays the actual behaviour given by the actual continuity support with the more distant member, depending on its greater or smaller flexural stiffness.

So, the bending moment decreases linearly starting from the value F dawn to the inversion of its action and to an opposite end value <F/2. Here, the diagram is deviated by the support reaction and oriented so to become null on the last end support of the beam.

The action of the loaded member may be transmitted to a system of several members converging into the connection joint, as indicated in Fig. 3.86a. In this

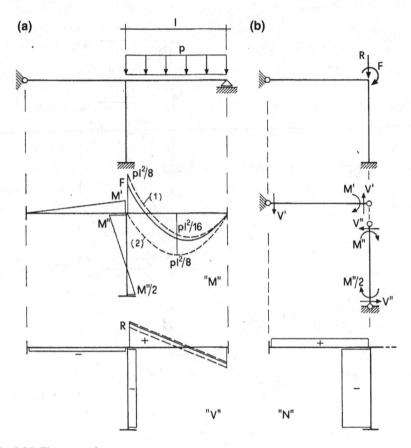

Fig. 3.86 Elementary frame system

case, placed the response between the two limit situations of full (1) or simple (2) support, the loaded member is replaced by its actions F ($<pl^2/8$) and R ($<5pl/8$ and $>pl/2$) as indicated in Fig. 3.86b. Subsequently, one proceeds with the *moment repartition* that, as already explained at Sect. 3.2.2, is made proportionally to the flexural stiffness of the converging members.

In order to define more easily the sense of the single actions, one can evidence the *equilibria at the connection joint* comparing "actions" and "reactions" as indicated in Fig. 3.87. Considering first the rotation equilibrium (see Fig. 3.87a), one opposes, to the clockwise action coming from the loaded member, the two reactions M' and M'' that shall be counterclockwise so to ensure the equilibrium $M' + M'' = F$. On the basis of the sense of the moments, one identifies the tensioned fibres of the involved members, while from the condition written above comes an indication on the intensity of the distributed moments that should be both smaller than the action F.

Fig. 3.87 Equilibria of the
connection joint

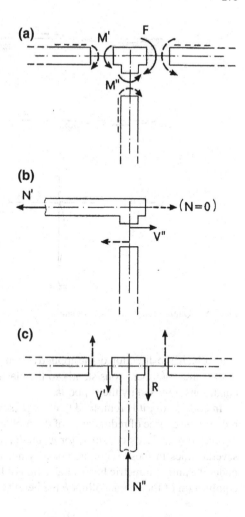

Starting from the values M' and M'', the bending moments of the two involved members have a linear variation, following the behaviour of elementary well-known situations. On the beam, it becomes null at its left end support; on the column, it inverts its effect dawn to a halved value at the fixed foot. Again in Fig. 3.86, the rotation equilibria of the quoted members are shown, from which negative (counter-clockwise) values of the shear force follow, consistent with the upward inclination of the corresponding bending moment diagrams.

For the diagrams of the axial force, one can refer to the translation horizontal and vertical equilibria represented in Fig. 3.87b, c that show the orthogonal relations of the shear forces V'' and $V' + R$ with the axial forces N' and N''.

The drawing of the qualitative diagrams of the internal forces can be also facilitated by *flexural deformed profile* deduced intuitively from the possible behaviour of the structure. As indicated in Fig. 3.88, one starts from the loaded member placing its

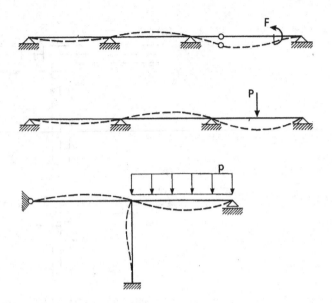

Fig. 3.88 Qualitative flexural deformation

deflection in the direction of the applied action. The deformation line is then extended
to the other members consistently to the elastic continuity of the profile and to the
conditions imposed by the supports.

In conclusion, it is reminded that the presence of *symmetries or anti-symmetries*
reduces the degree of redundancy of the problem and facilitates the estimation of the
internal forces. Let's consider, for example, a simple portal, like the one analysed
several times in these sections. For a symmetric arrangement of the structure and
under the anti-symmetric horizontal action of Fig. 3.89b, the condition of translation
equilibrium of the beam, calling h the height of the columns, is

$$\frac{M' + M''}{h} = V_0 = \frac{H}{2}$$

The ratio $r = M'/M''$ between the base and the top moments remains indeterminate,
depending on the stiffness k' and k'' of the different members. One can only estimate
the base moment of the columns greater than the top moment ($M' > M''$) due to the
rotation flexibility of the restraint given by the beam and to the opposite full restraint
of the base supports.

If the two extreme limit situations are considered, any redundancy disappears. For
a flexural null stiffness of the beam (Fig. 3.89a), the equilibrium translation equation:

$$\frac{M_0}{h} = V_0 = \frac{H}{2}$$

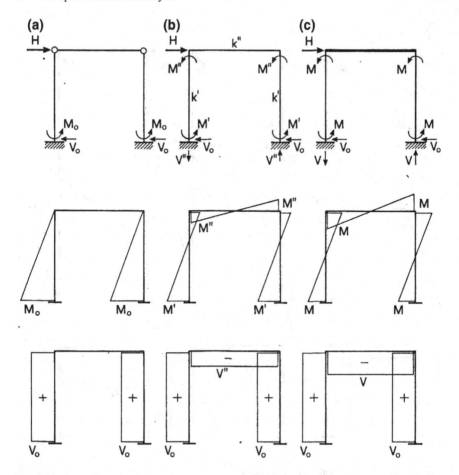

Fig. 3.89 Case of symmetric portal

provides directly the value of the moment at the base of the columns, independently from the elastic stiffness of the different members. For the opposite limit situation of perfectly rigid beam, the vertical anti-symmetry of the columns leads to the same value of the base and top moments, so that these latter can be directly obtained from:

$$\frac{2M}{h} = V_0 = \frac{H}{2}$$

The considerations above lead to the diagrams of bending moments drawn in Fig. 3.89. The diagrams of the shear force are defined on the basis of the values:

$$V_0 = \frac{H}{2} \quad \text{for the columns}$$

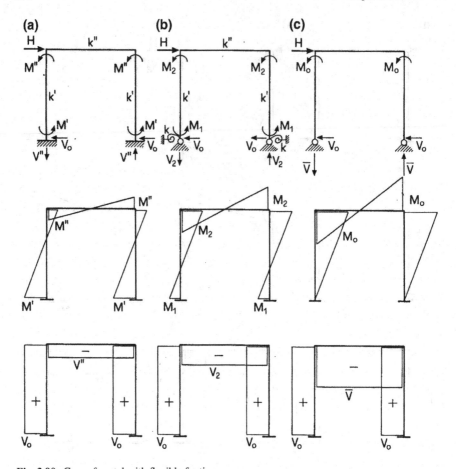

Fig. 3.90 Case of portal with flexible footings

$$V'' = -\frac{2M''}{l} \quad \text{or} \quad V = -\frac{2M}{l} \quad \text{for the beam}$$

where l is the length of the beam. Eventually, the axial forces are obtained from the vertical joint equilibria with the shear forces.

Considering now, starting from the same intermediate situation of Fig. 3.89b, an increasing rotation flexibility of the base supports, the diagrams modify as indicated in Fig. 3.90d. The ratio

$$r = \frac{M_1}{M_2} = r(k', k'', k)$$

between the base and the top moments decreases with the lower stiffness k of the base supports, while the top moments increase. For a null stiffness, with hinged base

supports (Fig. 3.90e), the situation becomes again statically determined and ruled by
the equilibrium

$$\frac{M_0}{h} = V_0 = \frac{H}{2}$$

that corresponds to that of the situation (a) overturned. In the last situation (e) examined, without rotation resistance of the base supports, the equilibrium is ensured
thanks to an increased flexural engagement of the beam.

Chapter 4
Second-Order Analysis

Abstract This chapter deals with the behaviour of slender members subjected to high axial forces. For these members, the deflections due to flexural loads lead to sensible eccentricities of the axial forces that add relevant contributions to the bending moments. These contributions cannot be any more neglected, as done within the first-order theory presented in the previous chapters. So the frame analysis shall move within a second-order theory as presented in what follows. Single members under eccentric tension or compression forces are firstly analysed, moving then to more complex structural assemblies, such as continuous beams or columns and frames, to which the criteria of the Force and Displacements Methods are extended, concluding with the presentation of the special calculation problems of the instability analysis of frame structures.

4.1 Second-Order Flexural Analysis of Beams

The flexural deformations of beams have been analysed at Sect. 2.1 in the case of absence of axial action. The results have been extended also to the case of presence of axial action assuming that this would have a small influence on the flexural behaviour of the beams. In this way, for example, the frame analysis, treated in detail in Chaps. 2 and 3, employed in an approximate way the elastic coefficients (flexibilities and stiffness) *of the first order*, strictly valid only for beams without axial action.

This approximation is no more valid when one has to deal with high axial actions in very flexible beams. In these cases, the transverse displacements of the beam axis due to the flexural loads are no more negligible, displaying sensible eccentricities of the axial action in the deformed configuration. So these axial actions can add relevant flexural contributions, called *of the second order*, as shown in detail in the following.

Let's consider the simply supported beam of Fig. 4.1 assumed with constant cross section. In addition to the flexural action F, a tensile axial action is present. The flexural action leads to the linear diagram of bending moment indicated in the figure. This diagram represents the first-order contribution with

$$M_1(x) = F\,x/l$$

© Springer Nature Switzerland AG 2019
G. Toniolo, *Introduction to Frame Analysis*, Springer Tracts
in Civil Engineering, https://doi.org/10.1007/978-3-030-14664-1_4

Fig. 4.1 Beam in flexure
and tension

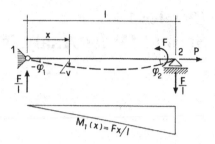

$$M_1(x) = Fx/l$$

In the same figure, the flexural deflection of the beam is indicated with the dashed line, showing for the current section of abscissa x an eccentricity v of the axial action. This axial action adds, to the preceding one, the second-order contribution

$$M_2(x) = -Pv(x)$$

that varies along the beam with the same law of the deflection line and has an opposite sign because it stretches the upper edge of the beam.

The equilibrium condition of the current section is set equalizing the elastic return moment to the moment due to the external actions:

$$EI\frac{d^2v}{dx^2} = -M(x)$$

With $M(x) = M_1(x) + M_2(x)$ one has:

$$EI\frac{d^2v}{dx^2} = -F\frac{x}{l} + Pv$$

Setting:

$$\alpha^2 = \frac{P}{EI}$$

and putting in the right order the terms, one obtains the differential equation:

$$v^{II} - \alpha^2 v = -\frac{F}{EI}\frac{x}{l}$$

The characteristic equation of the related homogeneous equation is:

$$\lambda^2 - \alpha^2 = 0$$

with the real roots:

$$\lambda_{1,2} = \pm\alpha$$

that lead to the function:

$$v_o(x) = B_1 e^{+\alpha x} + B_2 e^{-\alpha x} = C_1 \text{ch}\,\alpha x + C_2 \text{sh}\alpha x$$

The particular integral:

$$v_p(x) = ax^2 + bx + c$$

is now added that, put in the given equation, gives

$$2a - \alpha^2(ax^2 + bx + c) = -\frac{F}{lEI}x$$

Equalizing the coefficients of the terms of the same degree, one obtains:

$$a = 0 \quad c = 0$$

$$b = \frac{F}{\alpha^2 lEI}$$

and the solution becomes:

$$v(x) = C_1 \text{ch}\,\alpha x + C_2 \text{sh}\alpha x + \frac{F}{\alpha^2 lEI}x$$

Eventually, the boundary conditions are set corresponding to the end simple supports of the beam:

$$v(0) = 0$$

$$v(1) = 0$$

obtaining:

$$C_1 = 0$$

$$C_2 = -\frac{F}{\alpha^2 EI\text{sh}\,\alpha 1}$$

So the elastic line of the beam, inclusive of second-order effects, and its derivatives are:

$$v(x) = \frac{F}{\alpha^2 EI}\left\{\frac{1}{1}x - \frac{1}{\text{sh}\alpha l}\text{sh}\alpha x\right\}$$

$$-\varphi(x) = \frac{F}{\alpha\,EI}\left\{\frac{1}{\alpha l} - \frac{1}{sh\alpha l}ch\,\alpha\,x\right\}$$

$$M(x) = -EI\frac{d^2v}{dx^2} = \frac{F}{sh\alpha l}sh\alpha x$$

$$V(x) = -EI\frac{d^3v}{dx^3} = \frac{F}{l}\frac{\alpha l}{sh\alpha l}ch\,\alpha\,x$$

In particular, at the ends of the beam (with x = 0 and x = l), one has the rotations:

$$-\varphi_1 = \frac{Fl}{6EI}\left\{\frac{6}{\alpha l}\left(\frac{1}{\alpha l} - \frac{1}{sh\alpha l}\right)\right\}$$

$$\varphi_2 = \frac{Fl}{3EI}\left\{\frac{3}{\alpha l}\left(\frac{1}{th\alpha l} - \frac{1}{\alpha l}\right)\right\}$$

to which, with F = 1, correspond the "corrected" flexibilities, direct d_1 (=d_2) and indirect d_i:

$$d_1 = \frac{1}{3EI}f_1(\alpha l)$$

$$d_i = \frac{1}{6EI}f_i(\alpha l)$$

where

$$f_1(\alpha l) = \left\{\frac{3}{\alpha l}\left(\frac{1}{th\alpha l} - \frac{1}{\alpha l}\right)\right\}$$

$$f_i(\alpha l) = \left\{\frac{6}{\alpha l}\left(\frac{1}{\alpha l} - \frac{1}{sh\alpha l}\right)\right\}$$

are the *corrective functions* of the second order.

One notice that, without axial action ($\alpha = 0$), one has the same values found at Sect. 2.1.2, being:

$$\lim_{\alpha l\to 0} f_1(\alpha l) = 1$$

$$\lim_{\alpha l\to 0} f_i(\alpha l) = 1$$

while, for not null axial actions ($\alpha > 0$), one has reduction corrections with:

Fig. 4.2 Superposition of effects

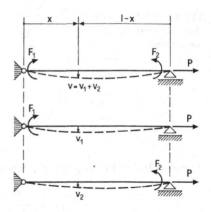

$$0 < f_1(\alpha l) < 1$$

$$0 < f_i(\alpha l) < 1$$

So the effect of the tensile axial force, added to the flexural action, is to reduce the deflection of the beam.

In Fig. 4.2 the same beam examined above is represented assuming that it is subjected to the two flexural actions F_1 and F_2. The first-order bending moment in this case is given by the function:

$$M_1(x) = F_2 \frac{x}{l} + F_1\left(1 - \frac{x}{l}\right)$$

and the equilibrium equation of the current section, again with $\alpha^2 = P/EI$, becomes:

$$v^{II} - \alpha^2 v = -\frac{F_2}{EI}\frac{x}{l} - \frac{F_1}{EI}\left(1 - \frac{x}{l}\right)$$

For such linear differential equation, the superposition of the effects of the different known terms can be applied and this is obviously valid for the same first member. So the effect of superposition shall be applied under the same axial force as shown in Fig. 4.2. Regarding the deformed line v, this force represents an intrinsic flexural characteristic of the beam, to be kept unchanged in all the single load conditions.

Using the preceding solution, for the case of concern, one can add the effects of the couple F_2 written for the section of abscissa x to the effects of the couple F_1 written for the section of abscissa $l - x$, obtaining:

$$v(x) = \frac{1}{\alpha^2 EI}\left\{F_2\left[\frac{x}{l} - \frac{sh\alpha x}{sh\alpha l}\right] + F_1\left[\frac{(l-x)}{l} - \frac{sh\alpha(l-x)}{sh\alpha l}\right]\right\}$$

Fig. 4.3 Bending moment diagrams for increasing slenderness

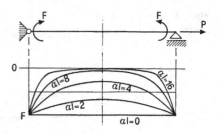

In particular for $F_1 = F_2 = F$, the function of the bending moment becomes:

$$M(x) = \frac{F}{\text{sh}\,\alpha\,l}[\text{sh}\,\alpha\,x + \text{sh}\,\alpha\,(1-x)]$$

with the minimum value at mid-span (for $x = 1/2$):

$$M_o(\alpha\,l) = F\frac{2\text{sh}\,\alpha l/2}{\text{sh}\,\alpha l} = F\frac{1}{\text{ch}\,\alpha\,l/2}$$

One can notice the correction factor that reduces the moment from the first-order value F (for $\alpha l = 0$) to smaller and smaller values ($M_o = 0$ for $\alpha l \rightarrow \infty$).

Figure 4.3 shows the diagrams of the bending moment for increasing values of αl. One can notice the progressive reduction, starting from the constant course, down to the peculiar situations of the ties, very slender and highly tensioned, on which the flexural actions decrease very rapidly within end segments very short with respect to the overall length the member.

For the resistance verification, since also the second-order theory assumes $\cos\varphi \approx 1$ and consequently the axial action remains practically constant along all the member ($N = P\cos\varphi \approx P$), the critical sections remain the end ones where the bending moment component M (=F) is maximum.

4.1.1 Eccentrically Loaded Columns

The case of eccentric axial compression (see Fig. 4.4) is analysed with the same approach of the preceding one: only the sign of the second-order term changes:

$$M_2(x) = +Pv(x)$$

Fig. 4.4 Beam in flexure and compression

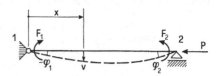

having the same orientation of the applied flexural action. For $F_1 = 0$ and $F_2 = F$, the equilibrium equation becomes:

$$v^{II} + \alpha^2 v = -\frac{F}{EI}\frac{x}{l}$$

where again $\alpha^2 = P/EI$. The characteristic equation of the related homogeneous equation leads to the imaginary roots:

$$\lambda_{1,2} = \pm i\,\alpha$$

So the solution has the circular functions instead of the hyperbolic ones:

$$v_0(x) = C_1\cos \alpha x + C_2\sin \alpha x$$

and the particular integral changes its sign

$$v_p(x) = -\frac{F}{\alpha^2 lEI}x$$

After the elaboration of the boundary conditions, corresponding to the end simple supports of the beam with $v(0) = 0$ and $v(l) = 0$, one obtains eventually the functions:

$$v(x) = \frac{F}{\alpha^2 EI}\left\{\frac{1}{\sin \alpha l}\sin \alpha x - \frac{1}{l}x\right\}$$

$$-\varphi(x) = \frac{F}{\alpha EI}\left\{\frac{1}{\sin \alpha l}\cos \alpha x - \frac{1}{\alpha l}\right\}$$

$$M(x) = -EI\frac{d^2 v}{dx^2} = \frac{F}{\sin \alpha l}\sin \alpha x$$

$$V(x) = -EI\frac{d^3 v}{dx^3} = \frac{F}{l}\frac{\alpha l}{\sin \alpha l}\cos \alpha x$$

In particular, at the ends of the beam (with $x = 0$ and $x = l$), one has the rotations:

$$-\varphi_1 = \frac{Fl}{6EI}\left\{\frac{6}{\alpha l}\left(\frac{1}{\sin \alpha l} - \frac{1}{\alpha l}\right)\right\}$$

$$\varphi_2 = \frac{Fl}{3EI}\left\{\frac{3}{\alpha l}\left(\frac{1}{\alpha l} - \frac{1}{tg\,\alpha l}\right)\right\}$$

to which, with $F = 1$, correspond the "corrected" flexibilities, direct d_1 (=d_2) and indirect d_i:

$$d_1 = \frac{1}{3EI} g_1(\alpha l)$$

$$d_i = \frac{1}{6EI} g_i(\alpha l)$$

where

$$g_1(\alpha l) = \left\{ \frac{3}{\alpha l} \left(\frac{1}{\alpha l} - \frac{1}{tg\alpha l} \right) \right\}$$

$$g_i(\alpha l) = \left\{ \frac{6}{\alpha l} \left(\frac{1}{\sin \alpha l} - \frac{1}{\alpha l} \right) \right\}$$

are the *corrective functions* of the second order or *instability functions* of the beam under flexure and compression.

Also for these functions one has

$$\lim_{\alpha l \to 0} g_1(\alpha l) = 1$$

$$\lim_{\alpha l \to 0} g_i(\alpha l) = 1$$

going back to the first-order flexibilities when the axial compression is null. But the circular functions at the denominator lead to infinite flexural effects under finite actions for some particular values of the axial compression. In fact one has:

$$g_1 = g_i = \infty \quad \text{for } \alpha l = n\,\pi$$

From this result, with n = 1, comes the definition of *critical load*, alternative to the static definition of indifferent flexural equilibrium given by Euler. Being $\alpha^2 = P/EI$, this new definition leads to the same value:

$$P_E = \frac{\pi^2 EI}{l^2}$$

and interprets the critical load as the value of the axial compression able to magnify to infinite the flexural effects of a finite action. From this interpretation, the equations will be deduced in the following text for the instability analysis of the frames with members in compression.

For axial actions lower than the critical load, that is for:

$$0 < \alpha l < \pi$$

Fig. 4.5 Effects of
increasing axial
compressions

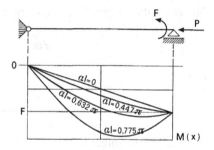

the values of the correction functions are greater than 1 ($g_1(\alpha l) > 1$ and $g_i(\alpha l) > 1$)
and ever-increasing as αl increases. So the axial compression increases the axial
flexibility, contrary to the axial tension that decreased it.

Similar considerations can refer to the internal forces, with bending moment
diagrams that, from the linear course corresponding to null axial force (see Fig. 4.5),
increase more and more up to curves with maximum value within the span of the
beam greater than the end value F. Therefore for the resistance verification, one has
to find previously the section of maximum moment.

Let's consider now the more general case of Fig. 4.4 where at the second end,
with $F_2 > 0$, the lower fibres are tensioned, and at the first end, with $-F_2 \le F_1 \le +$
F_2 the lower or the upper fibres may be tensioned. Similarly to the case of the beam
with tensile axial action, the solution can be set as:

$$v(x) = \frac{1}{\alpha^2 EI}\left\{F_2\left[\frac{\sin \alpha x}{\sin \alpha l} - \frac{x}{l}\right] + F_1\left[\frac{\sin \alpha(l - x)}{\sin \alpha l} - \frac{(l - x)}{l}\right]\right\}$$

and in terms of bending moment:

$$M(x) = \frac{1}{\sin \alpha l}\{F_2\sin \alpha x + F_1\sin \alpha(l - x)\} = \frac{F_2 - F_1\cos \alpha l}{\sin \alpha l}\sin \alpha x + F_1\cos \alpha x$$

where $\sin \alpha(l - x) = \sin \alpha l \cdot \cos \alpha x - \cos \alpha l \cdot \sin \alpha x$ has been put. So with

$$Z = \frac{F_2 - F_1\cos \alpha l}{F_1\sin \alpha l}$$

one has the equation of the bending moment:

$$M(x) = F_1\{Z\sin \alpha x + \cos \alpha x\}$$

and of its derivative:

$$\frac{dM}{dx} = \alpha F_1\{Z\cos \alpha x - \sin \alpha x\}$$

Therefore, the position of the maximum moment is given by:

$$Z\cos \alpha x - \sin \alpha x = 0$$

that is:

$$\text{tg}\, \alpha\, x = Z$$

that leads to the solution:

$$\alpha \overline{x} = \text{arctg} Z = \arcsin \frac{Z}{\sqrt{1+Z^2}} = \text{ar}\cos \frac{1}{\sqrt{1+Z^2}}$$

So for $\alpha \overline{x} < \alpha l$ one has a maximum moment $\overline{M} > F_2$ within the span of the beam, otherwise the critical section remains at its end 2 with $\overline{M} = F_2$. The limit situation with $\overline{x} = l$ is defined by:

$$\text{tg} \alpha l = Z = \frac{F_2 - F_1 \cos \alpha l}{F_1 \sin \alpha l}$$

that leads to:

$$F_1 = F_2 \cos \alpha l$$

For $F_1 < F_2 \cos \alpha l$ one has $\overline{x} > l$ and the maximum value of the bending moment remains $\overline{M} = F_2$. For $F_1 > F_2 \cos \alpha l$, one has $\overline{x} < l$ and the maximum bending moment should be calculated with:

$$\overline{M} = M(\alpha \overline{x}) = F_1 \left\{ \frac{Z^2}{\sqrt{1+Z^2}} + \frac{1}{\sqrt{1+Z^2}} \right\} = F_1 \sqrt{1+Z^2}$$

So, introducing the expression of Z, one has:

$$\overline{M} = \frac{1}{\sin \alpha l} \sqrt{F_2^2 + F_1^2 - 2F_2 F_1 \cos \alpha l}$$

In particular, for the case $F_1 = 0$ first treated, one has the maximum moment:

$$\overline{M} = \frac{1}{\sin \alpha l} F_2$$

to be assumed in place of F_2 when it is:

$$\cos \alpha l < 0$$

that is when the applied axial action is greater than ¼ of the Euler critical load:

$$\sqrt{\frac{P}{EI}}l > \frac{\pi}{2}$$

$$P > \frac{\pi^2 EI}{4l^2} = \frac{1}{4}P_E$$

In the anti-symmetric case with $F_1 = -F_2$, to have a maximum moment within the span of the beam, one should have:

$$\cos \alpha l < -1$$

which is impossible except at the limit $\alpha l = \pi$ of the critical load. For the resistance verification of the section, one can refer always to the end moments not affected by the second-order effects. The overall instability limit of the beam shall be separately verified in terms of $P < P_E$.

Much more sensible to the second-order effects are the situations that start from first-order distributions of moments close to the constant one. In fact, if it is $F_1 = F_2 = F$, the maximum moment is placed at the mid-span for symmetry reasons and is always greater than the applied end moments. Being $\sin \alpha l = 2 \cos \alpha l/2 \sin \alpha l/2$, one obtains:

$$\overline{M} = M(\alpha l/2) = \frac{1}{\cos \alpha l/2}F$$

Figure 4.6 shows the bending moment diagrams for some values of the ratio F_1/F_2 and for increasing values of αl, having omitted the case $F_1/F_2 = 0$ since it has been already shown in Fig. 4.5.

Instead of the quantity αl, as parameter that represents the different degrees of axial action, one can assume it's dimensionless value ν referred to the Euler critical load:

$$\alpha l = \sqrt{\frac{P}{EI}}l = \pi \sqrt{\frac{P}{\pi^2 EI/l^2}} = \pi \sqrt{\frac{P}{P_E}} = \pi \sqrt{\nu}$$

So, for example, the values chosen for the quantity αl of the diagram families of Figs. 4.5 and 4.6 correspond to the axial actions listed below:

$\alpha l/\pi$	$\nu = P/P_E$
0	0
0.447	0.2
0.632	0.4
0.775	0.6

Fig. 4.6 Effects of
increasing axial
compressions

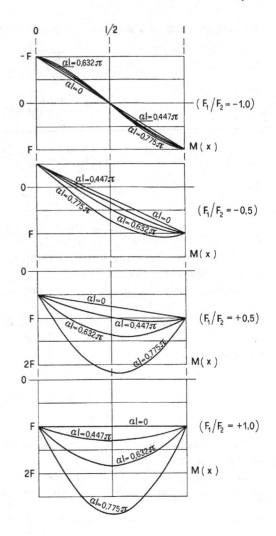

For what concerns the shear force, its variability:

$$V(x) = \frac{dM}{dx} = \alpha \left\{ \frac{F_2 - F_1 \cos \alpha l}{\sin \alpha l} \cos \alpha x - F_1 \sin \alpha x \right\}$$

with respect to the constant first-order value:

$$V_1(x) = \frac{F_2 - F_1}{1} = R$$

equal to the support reaction R, is due to the not negligible inclination of the deformed
beam axis. Assuming, in the field of the second-order theory:

Fig. 4.7 Second-order
effects on shear

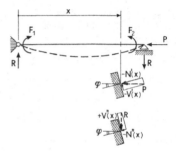

$$\sin\varphi = \varphi = \mathrm{tg}\varphi = -v^I(x) \ (\neq 0)$$

$$\cos\varphi \approx 1$$

on the current section acts a component of the action P transverse to the beam axis
equal to (see Fig. 4.7):

$$-P\sin\varphi = -P\varphi(x)$$

(positive if clockwise) to be added to the contribution of the reaction R:

$$+R\cos\varphi \approx +R$$

providing for the shear force:

$$V(x) = +R - P\varphi(x)$$

the same expression deduced above as derivative of the bending moment.

The reaction R has also an axial component, so that the axial force varies along
the beam with:

$$N(x) = -P\cos\varphi - R\sin\varphi = -P - R\varphi(x)$$

where

$$-\varphi(x) = \frac{1}{\alpha\,EI}\left\{\frac{F_2 - F_1\cos\alpha l}{\sin\alpha l}\cos\alpha x - F_1\sin\alpha x - \frac{F_2 - F_1}{\alpha\,l}\right\}$$

Figure 4.8 shows, only for the case $F_1 = 0$ and $F_2 = F$, the shear diagram family
for the same values of αl assumed in Fig. 4.5. These diagrams depend also from
the ratio R/P; with respect to the first-order value $N_1(x) = -P = $ const., there are

Fig. 4.8 Effects of
increasing axial compression
on shear

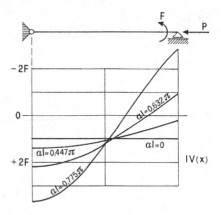

in general small variations so that one can omit them without sensible errors in the
resistance verifications of the sections.

4.1.2 Bending of Beams with Axial Force

For the beam of Fig. 4.9, the same analysis can be applied with a parabolic first-order
term:

$$M_1(x) = \frac{pl}{2}x - \frac{p}{2}x^2$$

so that the equilibrium equation becomes (with $\alpha^2 = P/EI$):

$$v^{II} + \alpha^2 v = -\frac{p}{2EI}(lx - x^2)$$

Assuming, for the particular integral, a second degree polynomial form, one has:

$$2a + \alpha^2(ax^2 + bx + c) = -\frac{P}{2EI}(lx - x^2)$$

Fig. 4.9 Beam with
distributed load

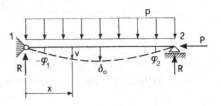

from which one obtains:

$$\alpha^2 a = +\frac{P}{2EI} \quad \text{with } a = +\frac{P}{2\alpha^2 EI}$$

$$\alpha^2 b = -\frac{pl}{2EI} \quad \text{with } b = -\frac{pl}{2\alpha^2 EI}$$

$$2a + \alpha^2 c = 0 \quad \text{with } c = -\frac{P}{\alpha^2 EI}$$

and the solution can be written as:

$$v(x) = C_1\cos \alpha x + C_2 \sin \alpha x + \frac{p}{2\alpha^2 EI}\left[x^2 - lx - \frac{2}{\alpha^2}\right]$$

From the first boundary condition, with $v(0) = 0$, one has:

$$C_1 = +\frac{p}{\alpha^4 EI}$$

and from the second, with $v(l) = 0$, one has:

$$\frac{p}{\alpha^4 EI}\cos \alpha l + C_2 \sin \alpha l - \frac{p}{\alpha^4 EI} = 0$$

that leads to:

$$C_2 = +\frac{p}{\alpha^4 EI}\frac{1 - \cos \alpha l}{\sin \alpha l}$$

So one obtains the functions:

$$v(x) = \frac{p}{\alpha^4 EI}\left\{\cos \alpha x + \frac{1 - \cos \alpha l}{\sin \alpha l}\sin \alpha x + \frac{\alpha^2}{2}x^2 - \frac{\alpha^2 l}{2}x - 1\right\}$$

$$-\varphi(x) = \frac{p}{\alpha^3 EI}\left\{-\sin \alpha x + \frac{1 - \cos \alpha l}{\sin \alpha l}\cos \alpha x + \alpha x - \frac{\alpha l}{2}\right\}$$

$$M(x) = -EI\frac{d^2 v}{dx^2} = \frac{p}{\alpha^2}\left\{\cos \alpha x + \frac{1 - \cos \alpha l}{\sin \alpha l}\sin \alpha x - 1\right\}$$

$$V(x) = -EI\frac{d^3 v}{dx^3} = \frac{p}{\alpha}\left\{-\sin \alpha x + \frac{1 - \cos \alpha l}{\sin \alpha l}\cos \alpha x\right\}$$

From this solution, the maximum deflection (at mid-span) is deduced as:

$$\delta_o = v(l/2) = \frac{5pl^4}{384EI}\left\{\frac{384}{5\alpha^4 l^4}\left[\frac{1}{\cos \alpha l/2} - \frac{\alpha^2 l^2}{8} - 1\right]\right\}$$

where the first-order expression is evidenced. The instability function placed in brackets tends to unit for $\alpha l \to 0$ and tends to infinite for $\cos \alpha l/2 \to 0$. This latter condition corresponds to

$$\frac{\alpha l}{2} = \frac{2n-1}{2}\pi$$

and, with $n = 1$, leads again to Euler critical load.

The end rotations on the supports are:

$$-\varphi_1 = \varphi_2 = \frac{pl^3}{24EI}g_o(\alpha l)$$

with the instability function:

$$g_o = \left\{\frac{24}{\alpha^3 l^3}\left[\frac{1-\cos\alpha l}{\sin\alpha l} - \frac{\alpha l}{2}\right]\right\}$$

placed again as amplification of the first-order contribution.

Eventually, one can deduce the value of the maximum (mid-span) moment:

$$\overline{M} = M(l/2) = \frac{pl^2}{8}\left\{\frac{8}{\alpha^2 l^2}\left[\frac{1}{\cos\alpha l/2} - 1\right]\right\}$$

and the value of the maximum shear on the supports:

$$V_1 = -V_2 = \frac{pl}{2}\left\{\frac{2}{\alpha l}\frac{1-\cos\alpha l}{\sin\alpha l}\right\} = R\left[1 - \frac{P}{R}\varphi_1\right]$$

with $R = pl/2$ equal to the support reaction of the beam (see Fig. 4.9).

It is reminded that the superposition of flexural effects can be applied under the same axial action; therefore, situations like that of Fig. 4.10 can be treated with simple additions of expressions as:

$$M(x) = \frac{X_j - X_i\cos\alpha l}{\sin\alpha l}\sin\alpha x + X_i\cos\alpha x + \frac{p}{\alpha^2}\left\{\cos\alpha x + \frac{1-\cos\alpha l}{\sin\alpha l}\sin\alpha x - 1\right\}$$

Fig. 4.10 Continuous beam

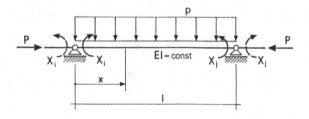

For completeness sake also the expressions related to the beam under tension are listed, omitting their detailed deduction that is based on elaborations quite similar to those reported here above.

$$v(x) = \frac{p}{\alpha^4 EI}\left\{ch\,\alpha\,x - \frac{ch\,\alpha\,1 - 1}{sh\alpha l}sh\alpha x - \frac{\alpha^2}{2}x^2 + \frac{\alpha^2 l}{2}x - 1\right\}$$

$$-\varphi(x) = \frac{p}{\alpha^3 EI}\left\{sh\alpha x - \frac{ch\,\alpha\,1 - 1}{sh\alpha l}ch\alpha x - \alpha x + \frac{\alpha l}{2}\right\}$$

$$M(x) = -EI\frac{d^2 v}{dx^2} = \frac{p}{\alpha^2}\left\{-ch\alpha x + \frac{ch\,\alpha\,1 - 1}{sh\alpha l}sh\alpha x + 1\right\}$$

$$V(x) = -EI\frac{d^3 v}{dx^3} = \frac{p}{\alpha}\left\{-sh\alpha x + \frac{ch\,\alpha\,1 - 1}{sh\alpha l}ch\alpha x\right\}$$

$$\delta_o = v(l/2) = \frac{5pl^4}{384EI}\left\{\frac{384}{5\alpha^4 l^4}\left[\frac{1}{ch\,\alpha\,l/2} + \frac{\alpha^2 l^2}{8} - 1\right]\right\}$$

$$-\varphi_1 = \varphi_2 = \frac{pl^3}{24EI}\left\{\frac{24}{\alpha^3 l^3}\left[\frac{\alpha l}{2} - \frac{ch\,\alpha\,1 - 1}{sh\alpha l}\right]\right\} = \frac{pl^3}{24EI}f_o(\alpha l)$$

$$\overline{M} = M(l/2) = \frac{pl^2}{8}\left\{\frac{8}{\alpha^2 l^2}\left[1 - \frac{1}{ch\,\alpha\,l/2}\right]\right\}$$

$$V_1 = -V_2 = \frac{pl}{2}\left\{\frac{2}{\alpha l}\frac{ch\,\alpha\,1 - 1}{sh\alpha l}\right\}$$

4.1.3   Cantilever Column

Another typical situation of compression element is represented by the column with a full support at its base and no restraints at the top as shown in Fig. 4.11. In this case, for a simpler solution, it is convenient to link the local reference system of axis x, y, z to the mobile top of the column, so that the second-order contribution, in the case of a load connected to the same top section, remains again represented by Pv:

$$EI\frac{d^2 v}{dx^2} + Pv = -F - Hx$$

Setting again $\alpha^2 = P/EI$ and keeping separated the solutions of the two flexural loads, with $F \neq 0$ and $H = 0$ one has the equation:

$$v^{II} + \alpha^2 v = -\frac{F}{EI}$$

that leads to the function:

Fig. 4.11 Cantilever column

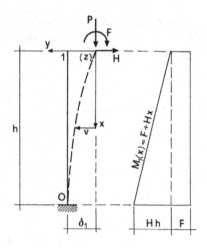

$$v = C_1 \cos \alpha x + C_2 \sin \alpha x - \frac{F}{\alpha^2 EI}$$

and to its first derivative:

$$-\varphi = -C_1 \alpha \sin \alpha x + C_2 \alpha \cos \alpha x$$

From the boundary conditions:

$$v(0) = 0$$
$$\varphi(h) = 0$$

one obtains the constants:

$$C_1 = \frac{F}{\alpha^2 EI}$$
$$C_2 = \frac{F}{\alpha^2 EI} \operatorname{tg} \alpha h$$

so that the solution is:

$$v(x) = \frac{F}{\alpha^2 EI}(\cos \alpha x + \operatorname{tg} \alpha h \sin \alpha x - 1)$$

$$-\varphi(x) = \frac{F}{\alpha EI}(-\sin \alpha x + \operatorname{tg} \alpha h \cos \alpha x)$$

$$M(x) = -EI\frac{d^2 v}{dx^2} = F(\cos \alpha x + \operatorname{tg} \alpha h \sin \alpha x)$$

$$V(x) = -EI\frac{d^3 v}{dx^3} = \alpha F(-\sin \alpha x + \operatorname{tg} \alpha h \cos \alpha x)$$

In particular, one obtains the maximum displacement (at the top) equal to:

$$\delta_1 = v(h) = \frac{Fh^2}{2EI}\left\{\frac{2}{\alpha^2 h^2}\left(\frac{1}{\cos\alpha h} - 1\right)\right\}$$

the maximum rotation (at the top) equal to:

$$-\varphi_1 = -\varphi(0) = \frac{Fh}{EI}\left(\frac{1}{\alpha h}\mathrm{tg}\alpha h\right)$$

and the maximum moment (at the base) equal to:

$$M_o = M(h) = F\frac{1}{\cos\alpha h}$$

In these equations, the first-order value is evidenced, modified always in increase by the pertinent instability functions. For $\cos\alpha h = 0$, one has infinite amplifications, so that the critical load is deduced with:

$$\overline{\alpha}h = \frac{\pi}{2}$$

that is:

$$P'_E = \frac{\pi^2 EI}{(2h)^2} = \frac{\pi^2 EI}{l_o^2}$$

having indicated with $l_o = \beta h$ the *buckling length* and with

$$\beta = \frac{\pi}{\overline{\alpha}h} = \frac{l_o}{h}$$

(=2 in the present case) the *supports factor* related to the instability behaviour of the column. So for the column of Fig. 4.11 one has, due to the lower supports degree, a buckling length double than that of the column with two end hinges and consequently, going with the inverse of exponent 2, one has a critical load four times lower.

Considering now the second load condition with $F = 0$ and $H \neq 0$, the following equation is set:

$$v^{II} + \alpha^2 v = -\frac{H}{EI}x$$

The solution is:

$$v = C_1\cos\alpha x + C_2\sin\alpha x - \frac{H}{\alpha^2 EI}x$$

and its first derivative is:

$$-\varphi = -C_1\alpha \sin \alpha x + C_2\alpha \cos \alpha x - \frac{H}{\alpha^2 EI}$$

From the boundary conditions $v(0) = 0$ and $\varphi(h) = 0$, one has:

$$C_1 = 0$$

$$C_2 = \frac{H}{\alpha^2 EI}\frac{1}{\cos \alpha h}$$

arriving to the following functions:

$$v(x) = \frac{H}{\alpha^2 EI}\left(\frac{1}{\cos \alpha h}\sin \alpha x - \alpha x\right)$$

$$-\varphi(x) = \frac{H}{\alpha EI}\left(\frac{1}{\cos \alpha h}\cos \alpha x - 1\right)$$

$$M(x) = -EI\frac{d^2 v}{dx^2} = \frac{H}{\alpha}\frac{\sin \alpha x}{\cos \alpha h}$$

$$V(x) = -EI\frac{d^3 v}{dx^3} = H\frac{\cos \alpha x}{\cos \alpha h}$$

In particular, one obtains (at the top) a maximum displacement equal to:

$$\delta_1 = v(h) = \frac{Hh^3}{3EI}\left\{\frac{3}{\alpha^3 h^3}(\text{tg}\alpha h - \alpha h)\right\}$$

a maximum rotation (at the top) equal to:

$$-\varphi_1 = -\varphi(0) = \frac{Hh^2}{2EI}\left\{\frac{2}{\alpha^2 h^2}\left(\frac{1}{\cos \alpha h} - 1\right)\right\}$$

and a maximum moment (at the base) equal to:

$$M_o = M(h) = Hh\frac{\text{tg}\alpha h}{\alpha h}$$

having evidenced again the first-order value, to be amplified by the related instability function. Obviously, dealing with a problem of perfect and indefinite elasticity, the different arrangement of the load doesn't modify the value of the critical load nor that of the related supports factor ($\beta = 2$).

On the contrary for different load arrangements, there are different amplifications of the flexural effects. For example, the first-order value (=F or =Hh) receives a larger amplification in the case of the couple F than in the case of the force H, being $(1/\cos\alpha h) > (\text{tg}\alpha h/\alpha h)$. Of these amplification factors, sometimes approximate expressions are used corresponding the first two terms of their series development with:

$$M_o = \frac{1}{\cos \alpha h} F \approx \frac{1}{1 - (\alpha h)^2/2} F = \frac{1}{1 - 5.00\nu} F$$

or with:

$$M_o = \frac{tg\alpha h}{\alpha h} Hh \approx \frac{1}{1 - (\alpha h)^2/3} Hh = \frac{1}{1 - 3.33\nu} Hh$$

having set $\nu = P/P_E$, $P_E = \pi^2 EI/h^2$ and $\pi^2 \approx 10$. A similar truncated series development, applied to the situation of doubly hinged column (see Fig. 4.4 of Sect. 4.1.1) with symmetric distribution of moments ($F_1 = F_2 = F$), leads to the expression:

$$\overline{M} = \frac{1}{\cos \alpha l/2} F \approx \frac{1}{1 - (\alpha l)^2/8} F = \frac{1}{1 - 1.25\nu} F$$

that shows (with $l = h$) the smaller influence of the axial action on the flexural component of the internal force of the most stressed section.

4.2 Analysis by Force Method

With the solution of the cases treated in the preceding Sect. 4.1, one has obtained the elastic coefficients related to the second-order behaviour of the members subjected to flexure and tensile or compressive axial action. In particular, the corrected flexibilities and end rotations due to transverse load have been defined for the situations summarized in Fig. 4.12 for which one has:

Fig. 4.12 Definition of flexibilities

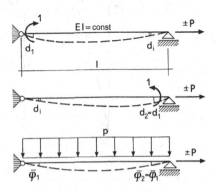

Tensioned members	Compressed members
$d_1 = \dfrac{1}{3EI} f_1$	$d_1 = \dfrac{1}{3EI} g_1$
$d_i = \dfrac{1}{6EI} f_i$	$d_i = \dfrac{1}{6EI} g_i$
$\varphi_1 = \dfrac{pl^3}{24EI} f_0$	$\varphi_1 = \dfrac{pl^3}{24EI} g_0$
$f_1 = \dfrac{3}{\alpha l}\left(\dfrac{1}{th\alpha l} - \dfrac{1}{\alpha l}\right)$	$g_1 = \dfrac{3}{\alpha l}\left(\dfrac{1}{\alpha l} - \dfrac{1}{tg\alpha l}\right)$
$f_i = \dfrac{6}{\alpha l}\left(\dfrac{1}{\alpha l} - \dfrac{1}{sh\alpha l}\right)$	$g_i = \dfrac{6}{\alpha l}\left(\dfrac{1}{sin\alpha l} - \dfrac{1}{\alpha l}\right)$
$f_0 = \dfrac{24}{(\alpha l)^3}\left(\dfrac{\alpha l}{2} - \dfrac{ch\alpha l - 1}{sh\alpha l}\right)$	$g_0 = \dfrac{24}{(\alpha l)^3}\left(\dfrac{1 - cos\alpha l}{sin\alpha l} - \dfrac{\alpha l}{2}\right)$

In the following pages, the tables with the numerical values of these functions in the ranges of possible use are reported.

On the basis of the flexibility expressions listed above and with the effect superposition as stated in the preceding section, it is possible to perform the second-order analysis following the same criteria of the Force Method.

Let's consider for example the case of Fig. 4.13a. The isostatic auxiliary structure is shown in Fig. 4.13b, where the effect of the suppressed rotation restraint is represented by the unknown moment X_1. The compatibility condition (see Fig. 4.13c, d):

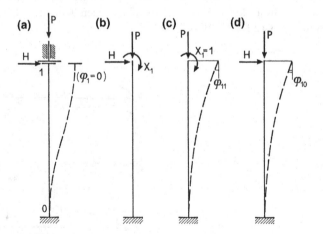

Fig. 4.13 Hyperstatic column

$$\varphi_{11}X_1 + \varphi_{10} = 0$$

expresses the annulment of the rotation at the end 1 of the member. The coefficients are deduced from the cases treated at Sect. 4.1.3:

$$\varphi_{11} = \frac{h}{EI}\left\{\frac{1}{\alpha h}\,\mathrm{tg}\,\alpha\,h\right\}$$

$$\varphi_{10} = \frac{Hh}{2EI}\left\{\frac{2}{(\alpha h)^2}\left(\frac{1}{\cos\alpha\,h} - 1\right)\right\}$$

Flexibilities of beams subjected to tension, second-order correction functions

αl	f_1	f_i	f_o
0.00	1.0000	1.0000	1.0000
0.25	0.9959	0.9927	0.9937
0.50	0.9837	9716	0.9756
0.75	0.9644	0.9381	0.9468
1.00	0.9391	0.8947	0.9092
1.25	0.9092	0.8436	0.8651
1.50	0.8762	0.7881	0.8167
1.75	0.8415	0.7305	0.7662
2.00	0.8000	0.6728	0.7152
2.25	0.7707	0.6167	0.6652
2.50	0.7363	0.5633	0.6170
2.75	0.7032	0.5133	0.5714
3.00	0.6716	0.4670	0.5288
3.25	0.6418	0.4247	0.4892
3.50	0.6138	0.3862	0.4526
3.75	0.5876	0.3514	0.4191
4.00	0.5630	0.3200	0.3885
4.25	0.5401	0.2919	0.3605
4.50	0.5187	0.2667	0.3350
4.75	0.4987	0.2441	0.3118
5.00	0.4801	0.2238	0.2906
5.25	0.4626	0.2057	0.2712
5.50	0.4463	0.1894	0.2536
5.75	0.4310	0.1748	0.2375
6.00	0.4167	0.1617	0.2228
6.25	0.4032	0.1499	0.2093
6.50	0.3905	0.1392	0.1969
6.75	0.3786	0.1296	0.1855
7.00	0.3673	0.1209	0.1751
7.25	0.3567	0.1130	0.1654

(continued)

(continued)

αl	f_1	f_i	f_o
7.50	0.3467	0.1058	0.1565
7.75	0.3371	0.0992	0.1483
8.00	0.3281	0.0932	0.1407
8.25	0.3196	0.0878	0.1336
8.50	0.3114	0.0828	0.1270
8.75	0.3037	0.0782	0.1209
9.00	0.2963	0.0739	0.1152
9.25	0.2893	0.0700	0.1099
9.50	0.2825	0.0664	0.1050
9.75	0.2761	0.0630	0.1003
10.00	0.2700	0.0599	0.0960
10.25	0.2641	0.0571	0.0919
10.50	0.2585	0.0544	0.0881
10.75	0.2531	0.0519	0.0845
11.00	0.2479	0.0496	0.0811
11.25	0.2430	0.0474	0.0780
11.50	0.2382	0.0454	0.0750
11.75	0.2336	0.0435	0.0721
12.00	0.2292	0.0417	0.0694

Flexibilities of beams subjected to compression, second-order correction functions

$\alpha l/\pi$	g_1	g_i	g_o
0.00	1.0000	1.0000	1.0000
0.04	1.0011	1.0019	1.0056
0.08	1.0042	1.0074	1.0061
0.12	1.0096	1.0168	1.0144
0.16	1.0173	1.0303	1.0260
0.20	1.0273	1.0481	1.0411
0.24	1.0401	1.0705	1.0603
0.28	1.0557	1.0983	1.0840
0.32	1.0746	1.1319	1.1126
0.36	1.0972	1.1724	1.1469
0.40	1.1241	1.2208	1.1880
0.44	1.1561	1.2787	1.2369
0.48	1.1941	1.3482	1.2954
0.52	1.2397	1.4318	1.3656
0.56	1.2946	1.5334	1.4507
0.60	1.3615	1.6582	1.5549
0.64	1.4442	1.8138	1.6843
0.68	1.5486	2.0117	1.8484
0.72	1.6836	2.2699	2.0616
0.76	1.8643	2.6185	2.3485

(continued)

(continued)

$\alpha l/\pi$	g_1	g_i	g_0
0.80	2.1179	3.1117	2.7530
0.84	2.4987	3.8579	3.3632
0.88	3.1333	5.1105	4.3842
0.92	4.4017	7.6292	6.4320
0.96	8.2038	15.2135	12.5865
1.00	$\pm\infty$	$\pm\infty$	$\pm\infty$
1.04	-6.9873	-15.2143	-12.0615
1.08	-3.1831	-7.6320	-5.9063
1.12	-1.9111	-5.1169	-3.8574
1.16	-1.2715	-3.8694	-2.8350
1.20	-0.8842	-3.1299	-2.2230
1.24	-0.6224	-2.6453	-1.8161
1.28	-0.4317	-2.3075	-1.5265
1.32	-0.2847	-2.0625	-1.3100
1.36	-0.1661	-1.8807	-1.1422
1.40	-0.0665	-1.7446	-1.0086
1.44	0.0201	-1.6434	-0.8997
1.48	0.0982	-1.5705	-0.8093
1.52	0.1711	-1.5221	-0.7332
1.56	0.2417	-1.4962	-0.6683
1.60	0.3127	-1.4926	-0.6122
1.64	0.3870	-1.5131	-0.5634
1.68	0.4684	-1.5618	-0.5205
1.72	0.5620	-1.6466	-0.4826
1.76	0.6759	-1.7815	-0.4487
1.80	0.8240	-1.9928	-0.4184
1.84	1.0338	-2.3341	-0.3910
1.88	1.3689	-2.9316	-0.3662
1.92	2.0195	-4.1647	-0.3436
1.96	3.9357	-7.9328	-0.3230
2.00	$+\infty$	$-\infty$	-0.3040

The solution is:

$$X_1 = -\frac{\varphi_{10}}{\varphi_{11}} = -\frac{Hh}{2}\left\{\frac{2}{\alpha h}\frac{1-\cos\alpha h}{\sin\alpha h}\right\}$$

and gives also, due to the anti-symmetry of the problem, the moment at the opposite end with $M_0 = -X_1$.

One can notice again the first-order value corrected by the amplifying function that inserts the effects of the axial load P. This function is equal to 1 for $\alpha h = 0$. For $\overline{\alpha}h = \pi$, one has the same value $\pi^2 EI/h^2$ of the critical load, equal to that P_E of the doubly hinged member, and the same value $\beta = \pi/\overline{\alpha}h = 1$ of the supports factor, with a buckling length equal to the length of the concerned member.

4.2.1   Beam with Single Redundancy

The member with one redundancy of Fig. 4.14a is analysed assuming the auxiliary structure of Fig. 4.14b and introducing in the expression of the compatibility equation:

$$\varphi_{11}X_1 + \varphi_{10} = 0$$

the pertinent second-order corrected flexibilities:

$$\varphi_{11} = \frac{1}{3EI}g_1$$

$$\varphi_{10} = -\frac{Fl}{6EI}g_i$$

One immediately obtains:

$$X_1 = -\frac{\varphi_{10}}{\varphi_{11}} = \frac{F}{2}\left\{\frac{g_i}{g_1}\right\}$$

and by consequence:

$$\varphi_2 = -\frac{Fl}{3EI}g_1 + \frac{1}{6EI}g_iX_1 = -\frac{Fl}{4EI}\left\{\frac{4g_1^2 - g_i^2}{3g_1}\right\}$$

For a member with a tension axial force, one would have in brackets the correction functions f_1, f_i instead of the instability functions g_1, g_i of the present case, remaining the solution identical for the rest.

The expressions obtained keep in evidence the first-order value and in brackets introduce the new composite correction functions. These latter are equal to 1 in the absence of axial action ($\alpha l = 0$) and increase more and more up to infinity for $g_1 = 0$. So the instability condition for the concerned member is:

$$\frac{3}{\alpha l}\left(\frac{1}{\alpha l} - \frac{1}{tg\alpha l}\right) = 0$$

Fig. 4.14 Member with one redundancy

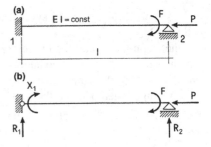

Fig. 4.15 Graphic
representation of tgαl = αl

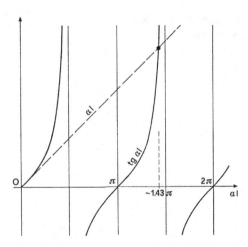

that is:

$$tg\alpha l = \alpha l \ (\alpha l > 0)$$

leading to the solution (see Fig. 4.15):

$$\bar{\alpha}l \simeq 1.43\pi$$

So one has a supports factor equal to:

$$\beta = \frac{\pi}{\alpha l} \approx 0.7$$

and a buckling length reduced to $l_o = 0.7l$, with a critical load equal to:

$$P'_E = \frac{\pi^2 EI}{l_o^2} = 2P_E$$

double than that of Euler's doubly hinged column.

For the other load conditions described in Fig. 4.16, one has the following solutions.

– Uniform distributed load p (Fig. 4.16a):

$$\varphi_{10} = \frac{pl^3}{24EI}g_o$$

from which:

Fig. 4.16 Different load conditions

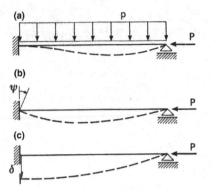

$$X_1 = -\frac{\varphi_{10}}{\varphi_{11}} = -\frac{pl^2}{8}\left\{\frac{g_o}{g_1}\right\}$$

$$\varphi_2 = +\frac{pl^3}{24EI}g_o + \frac{1}{6EI}g_iX_1 = \frac{pl^3}{48EI}\left\{g_o\frac{2g_1 - g_i}{g_1}\right\}$$

$$R_1 = \frac{pl}{2} - \frac{X_1}{1} = +\frac{5pl}{8}\left\{\frac{4g_1 + g_o}{5g_1}\right\}$$

$$R_2 = \frac{pl}{2} + \frac{X_1}{1} = +\frac{3pl}{8}\left\{\frac{4g_1 - g_o}{3g_1}\right\}$$

– Imposed rotation ψ of the fixed end (Fig. 4.16b):

$$\varphi_{10} = -\psi$$

from which:

$$X_1 = -\frac{\varphi_{10}}{\varphi_{11}} = +\frac{3EI}{1}\psi\left\{\frac{1}{g_1}\right\}$$

$$\varphi_2 = +\frac{1}{6EI}g_iX_1 = \frac{\psi}{2}\left\{\frac{g_i}{g_1}\right\}$$

$$R_2 = -R_1 = +\frac{X_1}{1} = \frac{3EI}{1^2}\psi\left\{\frac{1}{g_1}\right\}$$

– Imposed displacement δ of the fixed end (Fig. 4.16c):

$$\varphi_{10} = -\frac{\delta}{1}$$

from which:

Fig. 4.17 Displaced
configuration

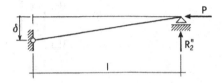

$$X_1 = -\frac{\varphi_{10}}{\varphi_{11}} = +\frac{3EI}{l^2}\delta\left\{\frac{1}{g_1}\right\}$$

$$\varphi_2 = +\frac{\delta}{l} + \frac{1}{6EI}g_i X_1 = \frac{3\delta}{2l}\left\{\frac{2g_1 + g_i}{3g_1}\right\}$$

For the evaluation of the reactions in this case, to the contribution $R'_2 = X_1/l$ of the flexural action the contribution of the axial force is to be added. In the displaced configuration (Fig. 4.17), the rotation equilibrium is:

$$P\delta + R''_2 l = 0$$

from which one obtains:

$$R''_2 = -\frac{P}{l}\delta$$

and the total reaction becomes:

$$R_2 = -R_1 = +\frac{X_1}{l} - \frac{P}{l}\delta = +\frac{3EI}{l^3}\delta\left\{\frac{1}{g_1} - \frac{(\alpha l)^2}{3}\right\}$$

It can be noticed how the direct translation stiffness, which is given by this last expression where $\delta = 1$ is set, is reduced in the second-order behaviour by the additional contribution of the compression axial action. On the contrary, for a tensile axial action, changing the sign of the term $P\delta$ in the rotation equilibrium, one has:

$$R_2 = -R_1 = +\frac{3EI}{l^3}\delta\left\{\frac{1}{f_1} + \frac{(\alpha l)^2}{3}\right\}$$

with an amplification effect on the direct translation stiffness of the member.

From the solutions provided above, for the member with a full and a hinged end supports, one obtains the elastic coefficients related to the Displacement Method that is the corrected stiffness for the second-order analysis. Setting $\psi = 1$ or $\delta = 1$ for the compressed member, one has:

$$k'_1 = \frac{3EI}{l}\left\{\frac{1}{g_1}\right\} = \frac{3EI}{l}G'(\alpha l)$$

$$k'_{vi} = \frac{3EI}{l^2}\left\{\frac{1}{g_1}\right\} = \frac{3EI}{l^2}G'(\alpha l)$$

$$k'_v = \frac{3EI}{l^3}\left\{\frac{1}{g_1} - \frac{(\alpha l)^2}{3}\right\} = \frac{3EI}{l^3}G'_v(\alpha l)$$

while for the tensioned member one has:

$$k'_1 = \frac{3EI}{l}\left\{\frac{1}{f_1}\right\} = \frac{3EI}{l}F'(\alpha l)$$

$$k'_{vi} = \frac{3EI}{l^2}\left\{\frac{1}{f_1}\right\} = \frac{3EI}{l^2}F'(\alpha l)$$

$$k'_v = \frac{3EI}{l^3}\left\{\frac{1}{f_1} + \frac{(\alpha l)^2}{3}\right\} = \frac{3EI}{l^3}F'_v(\alpha l)$$

For the fix end moment due to the load p, one has, respectively:

$$-m'_1 = \frac{pl^2}{8}\left\{\frac{g_0}{g_1}\right\} = \frac{pl^2}{8}G'_0(\alpha l)$$

$$-m'_1 = \frac{pl^2}{8}\left\{\frac{f_0}{f_1}\right\} = \frac{pl^2}{8}F'_0(\alpha l)$$

4.2.2   Beam with Double Redundancy

The member with two redundancies of Fig. 4.18a is analysed assuming the auxiliary structure of Fig. 4.18b. The compatibility conditions are:

$$\begin{cases} \varphi_{11}X_1 + \varphi_{12}X_2 + \varphi_{10} = 0 \\ \varphi_{21}X_1 + \varphi_{22}X_2 + \varphi_{20} = 0 \end{cases}$$

and, with:

Fig. 4.18 Member with two redundancies

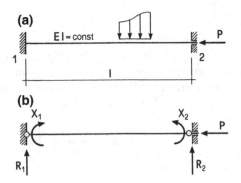

Fig. 4.19 Different load conditions

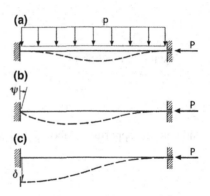

$$\varphi_{11} = \varphi_{22} = \frac{1}{3EI}g_1$$

$$\varphi_{12} = \varphi_{21} = \frac{1}{6EI}g_i$$

becomes:

$$\begin{cases} 2g_1X_1 + g_iX_2 = -\dfrac{6EI}{l}\varphi_{10} \\ g_iX_1 + 2g_1X_2 = -\dfrac{6EI}{l}\varphi_{20} \end{cases}$$

and leads to the solution:

$$X_1 = -\frac{6EI}{l}\frac{2g_1\varphi_{10} - g_i\varphi_{20}}{4gl^2 - gi^2}$$

$$X_2 = -\frac{6EI}{l}\frac{2g_1\varphi_{20} - g_i\varphi_{10}}{4g_1^2 - g_i^2}$$

This solution is here referred to the load conditions described in Fig. 4.19.

– Uniform distributed load p (Fig. 4.19a):

$$\varphi_{10} = \varphi_{20} = \frac{pl^3}{24EI}g_o$$

from which one obtains:

$$X_1 = X_2 = -\frac{pl^2}{4}\frac{2g_1 - g_i}{4g_1^2 - g_i^2}g_o = -\frac{pl^2}{12}\frac{3g_o}{2g_1 + g_i}$$

$$R_1 = +\frac{pl}{2} + \frac{X_2 - X_1}{1} = +\frac{pl}{2} = R_2$$

The instability condition is:

$$\frac{2g_1 + g_i}{g_o} = 0$$

and with the opportune elaborations becomes:

$$1 - \cos \alpha l = 0$$

from which one obtains (with $\alpha l > 0$):

$$\overline{\alpha}l = 2\pi$$

The supports factor of the double full restraint:

$$\beta = \frac{\pi}{\overline{\alpha}l} = 2\pi$$

indicates a halved buckling length ($l_o = 0.5l$) with a critical load:

$$P'_E = \frac{\pi^2 EI}{l_o^2} = 4P_E$$

four times greater than that of Euler's doubly hinged column.

– Imposed rotation ψ of a fixed end (Fig. 4.19b):

$$\varphi_{10} = -\psi$$
$$\varphi_{20} = 0$$

from which:

$$X_1 = +\frac{6EI}{1}\frac{2g_1}{4g_1^2 - g_i^2}\psi = +\frac{4EI}{1}\psi\left\{\frac{3g_1}{4g_1^2 - g_i^2}\right\}$$

$$X_2 = -\frac{6EI}{1}\frac{2g_i}{4g_1^2 - g_i^2}\psi = -\frac{2EI}{1}\psi\left\{\frac{3g_i}{4g_1^2 - g_i^2}\right\}$$

$$R_2 = -R_1 = +\frac{X_1 - X_2}{1} = +\frac{6EI}{1^2}\psi\left\{\frac{1}{2g_1 - g_i}\right\}$$

– Imposed displacement δ of a fixed end (Fig. 4.19c):

$$\varphi_{20} = -\varphi_{10} = \frac{\delta}{l}$$

from which

$$X_1 = -X_2 = +\frac{6EI}{l^2}\frac{2g_1 + g_i}{4g_1^2 + g_i^2}\delta = +\frac{6EI}{l^2}\delta\left\{\frac{1}{2g_1 - g_i}\right\}$$

$$R_2 = -R_1 = +\frac{X_1 - X_2}{l} - \frac{P}{l}\delta = +\frac{12EI}{l^3}\psi\left\{\frac{1}{2g_1 - g_i} - \frac{(\alpha l)^2}{12}\right\}$$

where one can notice the decrement contribution of the axial force, the same already deduced for the member with one redundancy (see Fig. 4.17). In this latter case, as the ratio to the first-order translation stiffness, this contribution is less relevant (four times smaller).

From the solutions provided above, for the member with two full end supports, one obtains the expressions of the corrected stiffness for the second-order analysis. Setting $\psi = 1$ or $\delta = 1$ for the compressed member, one has:

$$k_1 = \frac{4EI}{l}\left\{\frac{3g_1}{4g_1^2 - g_i^2}\right\} = \frac{4EI}{l}G_1(\alpha l)$$

$$k_i = \frac{2EI}{l}\left\{\frac{3g_i}{4g_1^2 - g_i^2}\right\} = \frac{2EI}{l}G_i(\alpha l)$$

$$k_{vi} = \frac{6EI}{l^2}\left\{\frac{1}{2g_1 - g_i}\right\} = \frac{6EI}{l^2}G_{vi}(\alpha l)$$

$$k_v = \frac{12EI}{l^3}\left\{\frac{1}{2g_1 - g_i} - \frac{(\alpha l)^2}{12}\right\} = \frac{12EI}{l^3}G_v(\alpha l)$$

while for the tensioned member one has:

$$k_1 = \frac{4EI}{l}\left\{\frac{3f_1}{4f_1^2 - f_i^2}\right\} = \frac{4EI}{l}F_1(\alpha l)$$

$$k_i = \frac{2EI}{l}\left\{\frac{3f_i}{4f_1^2 - f_i^2}\right\} = \frac{2EI}{l}F_i(\alpha l)$$

$$k_{vi} = \frac{6EI}{l^2}\left\{\frac{1}{2f_1 - f_i}\right\} = \frac{6EI}{l^2}F_{vi}(\alpha l)$$

$$k_v = \frac{12EI}{l^3}\left\{\frac{1}{2f_1 - f_i} - \frac{(\alpha l)^2}{12}\right\} = \frac{12EI}{l^3}F_v(\alpha l)$$

Fig. 4.20 Continuous two
storeys column

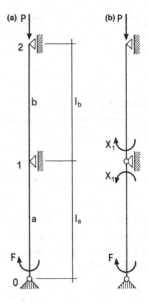

For the fix end moments due to the load p, one has, respectively:

$$-\overline{m}_1 = +\overline{m}_2 = \frac{pl^2}{12}\left\{\frac{3g_o}{2g_1 + g_i}\right\} = \frac{pl^2}{12}G_o(\alpha l)$$

$$-\overline{m}_1 = +\overline{m}_2 = \frac{pl^2}{12}\left\{\frac{3f_o}{2f_1 + f_i}\right\} = \frac{pl^2}{12}F_o(\alpha l)$$

4.2.3   Continuous Beams

The second-order analysis can be extended to the continuous beams following the
same criteria of the Force Method as treated at Sect. 2.2, using in the formulation of
the compatibility equations the corrected flexibilities. Let's consider for example the
simple case of Fig. 4.20a that can be solved with only one unknown: the continuity
moment X_1 on the intermediate support (see Fig. 4.20):

$$\left(\frac{l_a}{3EI_a}g_{1a} + \frac{l_b}{3EI_b}g_{1b}\right)X_1 + \frac{l_a}{6EI_a}g_{ia}F = 0$$

where

$$g_{1a} = g_1(\alpha_a l_a)$$
$$g_{ia} = g_i(\alpha_a l_a)$$

$$g_{lb} = g_1(\alpha_b l_b)$$

with

$$\alpha_a^2 = \frac{N_a}{EI_a}$$

$$\alpha_b^2 = \frac{N_b}{EI_b}$$

In the present case, the axial actions in the two members are the same:

$$N_a = N_b = N$$

The solution is obtained immediately with:

$$X_1 = \frac{-\frac{l_a}{6EI_a}g_{ia}}{\frac{l_a}{3EI_a}g_{1a} + \frac{l_b}{3EI_b}g_{1b}} :$$

After its numerical elaboration, this solution can be used for the analysis of any single member with the superposition of flexural effects quoted at Sect. 4.1.2 (see Fig. 4.10).

For the present case, a constant cross section is assumed with:

$$I_a = I_b = I$$

and:

$$\alpha_a = \alpha_b = \alpha = \sqrt{\frac{P}{EI}}$$

Setting:

$$l = \frac{l_a + l_b}{2}$$

the mean column length, and (with $l_a > l_b$):

$$\Delta l = \frac{l_a - l_b}{2} = cl$$

the semi-difference expressed as the ratio to the mean length through the factor:

$$c = \frac{\Delta l}{l}$$

substituting:

$$l_a = (1+c)l$$
$$l_b = (1-c)l$$

and introducing the pertinent simplifications, the solution becomes:

$$-X_1 = \frac{1}{2} \frac{(1+c)g_{ia}}{(1+c)g_{1a}+(1-c)g_{1b}}$$

In order to perform an instability calculation, one sets the condition:

$$(1+c)g_{1a}+(1-c)g_{1b}=0$$

and from this transcendental equation the root.

$$\alpha l = \overline{\alpha} l$$

is to be deduced to obtain the critical load:

$$P'_E = \alpha^2 EI$$

that corresponds to the collapse limit of the column.

Introducing the expressions of the instability functions, one has the subsequent steps here under reported:

$$\frac{3}{\alpha l}\left[\frac{1}{(1+c)\alpha l} - \frac{1}{tg(1+c)\alpha l}\right] + \frac{3}{\alpha l}\left[\frac{1}{(1-c)\alpha l} - \frac{1}{tg(1-c)\alpha l}\right] = 0$$

$$\frac{2}{(1-c^2)\alpha l} - \frac{tg(1-c)\alpha l + tg(1+c)\alpha l}{tg(1-c)\alpha l \cdot tg(1+c)\alpha l} = 0$$

$$\frac{2}{(1-c^2)\alpha l} = \frac{\cos(1+c)\alpha l \cdot \sin(1-c)\alpha l + \cos(1-c)\alpha l \cdot \sin(1+c)\alpha l}{\sin(1-c)\alpha l \cdot \sin(1+c)\alpha l}$$

$$\frac{2}{(1-c^2)\alpha l} = \frac{2\sin 2\alpha l}{\cos 2c\alpha l - \cos 2\alpha l}$$

and eventually:

$$\cos 2c\alpha l = (1-c^2)\,\alpha l\,\sin 2\alpha l + \cos 2\alpha l$$

So a simplified expression is obtained that enables to deduce some solutions. For equal column length ($l_a = l_b = l$) with $c = 0$, one has:

$$\overline{\alpha} l = \pi$$

Fig. 4.21 Graphic representation of the solutions

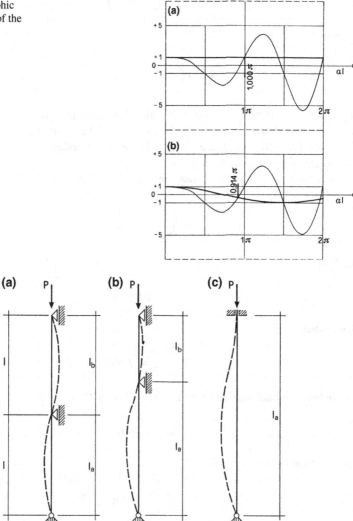

Fig. 4.22 Buckling shapes for different length ratios

that corresponds to the intersection of the first member curve with the second member curve drawn in Fig. 4.21a. From this solution, a supports factor

$$\beta = \frac{\pi}{\alpha l} = 1.0$$

is obtained that indicates a buckling length equal to the height of the column spans (see Fig. 4.22a):

$$l_o = l_a = l_b = l$$

with a critical load:

$$P'_E = \frac{\pi^2 EI}{l_o^2} = P_E$$

equal to what can be calculated with Euler's equation for any single column span separately. So in this case the equal flexibility of the columns, together with the constant value of the axial force, leads to the instability collapse limit without mutual actions exchanged between the parts.

For a column height double than the other ($l_a = 2l_b$), with $c = 0.333$ and:

$$l_a = 1.333l$$
$$l_b = 0.667l$$

one has:

$$\overline{\alpha}l = 0.914\pi$$

from which:

$$\beta = \frac{\pi}{\overline{\alpha}l} = 1.094$$

that indicates a buckling length equal to (see Fig. 4.22b):

$$l_o = 1.094l$$

and a critical load:

$$P'_E = \frac{\pi^2 EI}{l_o^2} = 0.836P_E$$

lower than what could be calculated with Euler's equation on the basis of the mean column length. If the stability verifications are referred to the single columns, setting:

$$l_o = \frac{\beta}{(1+c)}l_a = \beta_a l_a \quad \text{or}$$
$$l_o = \frac{\beta}{(1-c)}l_b = \beta_b l_b$$

one has:

$$\beta_a = \frac{1.094}{1.333} = 0.821 \quad \text{or}$$

$$\beta_b = \frac{1.094}{1.667} = 1.641$$

These values of the related supports factors show that, for the same axial force, the stiffer column b improves the stability of the contiguous column, and on the contrary, the more flexible column a worsens the stability of the former:

$$P'_E = \frac{1}{\beta_a^2} \frac{\pi^2 EI}{l_a^2} = 1.485 P_{Ea} \quad \text{or}$$

$$P'_E = \frac{1}{\beta_b^2} \frac{\pi^2 EI}{l_b^2} = 0.372 P_{Eb}$$

For a column much longer than the other, at the limit $l_b = 0$ with:

$$c = 1.0$$
$$l_a = 2.01$$

the instability condition, in the simplified form written above, becomes indeterminate. Going back to the original form, one has the equation:

$$g_{1a} = g_1(\alpha l_a) = 0$$

which is equal to the one of the member with one full and one hinged end supports (see Fig. 4.22c). As elaborated at Sect. 4.2.1, the solution gives (see Fig. 4.15):

$$\overline{\alpha} l = 0.143 \pi \ (= 2\alpha l)$$

from which

$$\beta_a = 0.7 \ (\beta = 2 \beta_a = 1.4)$$

So in conclusion the instability verification of the continuous constant section column of Fig. 4.20a can be performed with reference to the longer segment using the opportune supports factor β_a. This factor is included within the values 1.0 and 0.7 related to the extreme situations of equal and very different lengths. For intermediate situations, β_a should be evaluated as a function of the ratio c defined before, solving the pertinent equation as shown above.

Figure 4.23 shows the curves $\beta = \beta(c)$, $\beta_a = \beta_a(c)$ and $\beta_b = \beta_b(c)$. In particular, it can be seen how with good approximation a bilinear form for $1/\beta$ can be assumed:

$$1/\beta = 1.0 = \text{const.} \quad \text{for } c < 0.047$$
$$1/\beta = 1.014 - 0.30c \quad \text{for } c > 0.047$$

Fig. 4.23 Curves of support factors

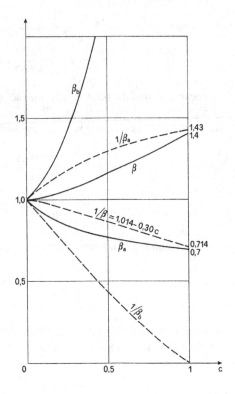

From this expression, the value of β can be deduced and subsequently $\beta_a = \beta/(1 + c)$ can be calculated for the verification:

$$P'_E = \frac{1}{\beta_a^2} \frac{\pi^2 EI}{l_a^2}$$

A similar analysis can be performed in the case of variation of an other of the parameters related to the flexibility of the column subjected to bending and axial compression. Let's consider the example of continuous column shown in Fig: 4.24 where the section remains constant and the lengths of the two segments are equal, while the axial actions vary with:

$$N_b = \frac{p_b s}{2} = P_b$$

$$N_a = N_b + \frac{p_a s}{2} = P_b + P_a$$

Setting $N_a = P$, and $N_b = c^2 P$ with $c^2 = N_b/N_a$, one has:

$$\alpha_a = \alpha = \sqrt{\frac{P}{EI}}$$

Fig. 4.24 Two-storey column

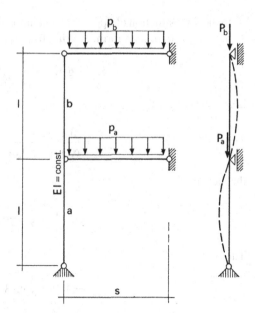

$$\alpha_b = c\sqrt{\frac{P}{EI}} = c\,\alpha$$

The instability condition $g_{1a} + g_{1b} = 0$ becomes:

$$\frac{3}{\alpha l}\left[\frac{1}{\alpha l} - \frac{1}{tg\alpha l}\right] + \frac{3}{c\,\alpha l}\left[\frac{1}{c\,\alpha l} - \frac{1}{tgc\,\alpha l}\right] = 0$$

and, after the opportune simplifications, leads to the expression:

$$(c^2 + 1)\sin\alpha l \cdot \sin c\alpha l = c^2\alpha l \sin(1 + c)\alpha l$$

For the limit situation of equal axial actions ($N_b = N_a$), with $c = 1$, one has the same equation:

$$1 = \alpha l \sin 2\alpha l + \cos 2\alpha l$$

from which, in the preceding case of Fig. 4.20 with $l_a = l_b = 1$, one obtained (see Fig. 4.22a).

$$\overline{\alpha}l = \pi \ \text{with}\,\beta = 1.0:$$

In the limit situation of upper segment unloaded ($N_b = 0$), setting $c = 0$ one would obtain an indeterminate form. Starting from the original instability condition where

$$g_{1b} = g_1(\alpha_b l) = g_1(0) = 1.0$$

is set, one obtains:

$$g_{1a} = \frac{3}{\alpha l}\left[\frac{1}{\alpha l} - \frac{1}{tg\alpha l}\right] = -1.0$$

that becomes:

$$tg\alpha l = \frac{3\alpha l}{3 + (\alpha l)^2}$$

The solution:

$$\overline{\alpha}l = 1.19\pi \text{ with } \beta = 0.84$$

corresponds to the maximum restraint degree given by the upper column segment to the lower more loaded one with reference to its instability.

For the intermediate situations, with $0 < c^2 < 1$, one has increasing values of β that can be deduced from the general form of the instability condition through the related solutions $\overline{\alpha}l$ ($\beta = \pi/\overline{\alpha}l$). The critical load varies consistently, decreasing from its maximum value:

$$P'_E = \frac{1}{0.84^2}\frac{\pi^2 EI}{l^2} = 1.416 P_E \text{ for } N_b = 0$$

dawn to the lower limit:

Fig. 4.25 Curve of the supports factor

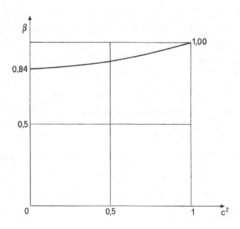

$$P'_E = P_E \text{ for } N_b = N_a$$

Figure 4.25 shows the related curve $\beta = \beta(c)$.

4.3 Analysis by Displacement Method

With the solution of the hyperstatic beams in the preceding section, the expressions have been provided of the corrected flexural stiffness, indicated with G or F, respectively, for the compressed or tensioned members. These stiffness can be used for the second-order analysis of frames following the criteria of the Displacement Method. For this analysis, the specific problems shall be considered related to the evaluation of the correction functions quoted above. These functions depend on the axial forces as distributed on the members of the frames that is on the applied loads and the related hyperstatics.

Same considerations are stated beforehand on the required numerical elaborations, noticing first that all the functions $F(\alpha l)$ and $G(\alpha l)$ become indeterminate for $\alpha l = 0$. Other indeterminate situations arise for the instability functions $G(\alpha l)$ for some values of the variable αl (for example $\pi/2$, π, ...), leading to forms such as 0/0.

In order to overcome these indeterminate forms, the expressions of the stiffness, originally set in terms of the flexibilities, shall be properly rearranged so to be numerically elaborated through their own algorithms that remain stable over all the range of possible use. Let's consider for example the expression:

$$G_1(\alpha l) = \frac{3g_1}{4g_1^2 - g_i^2}$$

of the correction function of the direct rotation stiffness. With the pertinent substitutions, this expression becomes:

$$G_1 = \frac{3\left[\dfrac{3}{\alpha l}\left(\dfrac{1}{\alpha l} - \dfrac{1}{\operatorname{tg}\alpha l}\right)\right]}{4\left[\dfrac{3}{\alpha l}\left(\dfrac{1}{\alpha l} - \dfrac{1}{\operatorname{tg}\alpha l}\right)\right]^2 - \left[\dfrac{6}{\alpha l}\left(\dfrac{1}{\sin \alpha l} - \dfrac{1}{\alpha l}\right)\right]^2}$$

and cannot be elaborated for $\alpha l = 0$, $\pi/2$, π, $3\pi/2$. With the possible simplifications, one obtains:

$$G_1 = \frac{\alpha l}{4} \frac{\sin \alpha l - \alpha l \cos \alpha l}{2 - 2\cos \alpha l - \alpha l \sin \alpha l}$$

that remains indeterminate only for $\alpha l = 0$. Let's consider then the series developments with which the circular functions are calculated:

$$\cos x = \sum_{n=0}^{\infty} \frac{(-1)^n}{(2n)!} x^{2n}$$

$$\sin x = \sum_{n=0}^{\infty} \frac{(-1)^n}{(2n+1)!} x^{2n+1}$$

For the numerator of the expression of G_1, one has (with $x = \alpha l$):

$$\sin x - x \cos x = +x - \frac{x^3}{3!} + \frac{x^5}{5!} - \frac{x^7}{7!} + \ldots - x + \frac{x^3}{2!} - \frac{x^5}{4!} + \frac{x^7}{6!} - \ldots$$

that is:

$$\sin x - x \cos x = +0 + \frac{2x^3}{3!} - \frac{4x^5}{5!} + \frac{6x^7}{7!} - \ldots = x^3 \sum_{n=1}^{\infty} \frac{-2n}{(2n+1)!}(-1)^n x^{2n-2}$$

For the denominator, one has:

$$2 - 2\cos x - x \sin x = \frac{2x^2}{2!} - \frac{2x^4}{4!} + \frac{2x^6}{6!} - \ldots + \frac{x^2}{1!} + \frac{x^4}{3!} - \frac{x^6}{5!} + \ldots$$

that is:

$$2 - 2\cos - x \sin x = 0 + \frac{2x^4}{4!} - \frac{4x^6}{6!} + \ldots = x^4 \sum_{n=2}^{\infty} \frac{2n-2}{(2n)!}(-1)^n x^{2n-4}$$

So setting, with $x = \alpha l$:

$$\kappa = (\alpha l)^2 = P l^2 / EI$$

where P is negative if compressive and the quantity κ has its sign, one obtains with simple substitutions in the expressions written above:

$$G_1(\kappa) = \frac{1}{4S(\kappa)} \sum_{n=1}^{\infty} \frac{2n}{(2n+1)!} \kappa^{n-1}$$

where

$$S(\kappa) = \sum_{n=2}^{\infty} \frac{2n-2}{(2n)!} \kappa^{n-2}$$

Through a similar procedure for the correction function of the members subjected to tension, one has:

$$F_1 = \frac{\alpha l}{4} \frac{\alpha lc\, h\alpha l - sh\alpha l}{2 - 2c\, h\alpha l - \alpha lsh\alpha l}$$

Also this expression becomes indeterminate for $\alpha l = 0$. So one uses the series developments of the hyperbolic functions:

$$c\, hx = \sum_{n=0}^{\infty} \frac{1}{(2n)!} x^{2n}$$

$$s\, hx = \sum_{n=0}^{\infty} \frac{1}{(2n+1)!} x^{2n+1}$$

obtaining, through the same passages above shown in detail, an equal expression:

$$F_1(\kappa) = \frac{1}{4S(\kappa)} \sum_{n=1}^{\infty} \frac{2n}{(2n+1)!} \kappa^{n-1} = G_1(\kappa)$$

to be elaborated, in this case of tensioned members, for positive values of the quantity

$$\kappa = (\alpha l)^2 = P l^2 / EI$$

So the numerical algorithm remains the same for compressed and tensioned members.

Stiffness of beams subjected to tension, second-order correction functions

αl	F_1	F_i	F_{vi}	F_v	F_o
0.00	1.0000	1.0000	1.0000	1.0000	1.0000
0.25	1.0021	0.9990	1.0010	1.0062	0.9996
0.50	1.0083	0.9959	1.0042	1.0250	0.9959
0.75	1.0186	0.9908	1.0093	1.0562	0.9907
1.00	1.0329	0.9838	1.0165	1.0999	0.9837
1.25	1.0511	0.9752	1.0258	1.1500	0.9749
1.50	1.0729	0.9650	1.0369	1.2244	0.9644
1.75	1.0982	0.9534	1.0500	1.3052	0.9524
2.00	1.1269	0.9407	1.0648	1.3982	0.9391
2.25	1.1586	0.9272	1.0815	1.5034	0.9247
2.50	1.1933	0.9129	1.0998	1.6207	0.9092
2.75	1.2305	0.8982	1.1198	1.7500	0.8931
3.00	1.2702	0.8832	1.1412	1.8912	0.8762
3.25	1.3121	0.8682	1.1641	2.0444	0.8590
3.50	1.3561	0.8531	1.1884	2.2093	0.8415

(continued)

(continued)

αl	F_1	F_i	F_{vi}	F_v	F_o
3.75	1.4018	0.8383	1.2140	2.3859	0.8237
4.00	1.4492	0.8238	1.2407	2.5741	0.8060
4.25	1.4981	0.8097	1.2686	2.7738	0.7883
4.50	1.5483	0.7900	1.2975	2.9850	0.7707
4.75	1.5997	0.7829	1.3274	3.2976	0.7533
5.00	1.6521	0.7703	1.3582	3.4414	0.7363
5.25	1.7055	0.7583	1.3898	3.6866	0.7195
5.50	1.7597	0.7469	1.4221	3.9430	0.7032
5.75	1.8147	0.7361	1.4552	4.2104	0.6872
6.00	1.8704	0.7259	1.4889	4.4889	0.6716
6.25	1.2967	0.7163	1.5232	4.7784	0.6565
6.50	1.9835	0.7072	1.5580	5.0789	0.6418
6.75	2.0408	0.6986	1.5934	5.3902	0.0276
7.00	2.0985	0.6906	1.6292	5.7125	0.6138
7.25	2.1566	0.6830	1.6654	6.0456	0.6005
7.50	2.2150	0.6759	1.7020	6.3895	0.5876
7.75	2.2738	0.6692	1.7389	6.7441	0.5751
8.00	2.3328	0.6629	1.7762	7.1095	0.5630
8.25	2.3921	0.6570	1.8137	7.4856	0.5513
8.50	2.4516	0.6515	1.8516	7.8724	0.5101
8.75	2.5113	0.6403	1.8897	8.2099	0.5292
9.00	2.5712	0.6414	1.9280	8.6780	0.5187
9.25	2.6313	0.6368	1.9665	9.0967	0.5085
9.50	2.6916	0.6324	2.0052	9.5260	0.4987
9.75	2.7519	0.6283	2.0441	9.9659	0.4892
10.00	2.8124	0.6244	2.0831	10.4164	0.4801
10.25	2.8731	0.6208	2.1223	10.8775	0.4712
10.50	2.9338	0.6173	2.1616	11.3491	0.4626
10.75	2.9946	0.6140	2.2011	11.8313	0.4543
11.00	3.0555	0.6109	2.2407	12.3240	0.4463
11.25	3.1165	0.6079	2.2803	12.8272	0.4385
11.50	3.1776	0.6051	2.3201	13.3410	0.4310
11.75	3.2388	0.6025	2.3600	13.8652	0.4237
12.00	3.3000	0.5999	2.4000	14.4000	0.4167
12.25	3.4389	0.5946	2.4908	15.6504	0.4015

Stiffness of beams subjected to compression, second-order correction functions

αl/π	G_1	G_i	G_{vi}	G_v	G_o
0.00	1.0000	1.0000	1.0000	1.0000	1.0000
0.04	0.9995	1.0003	0.9997	0.9984	1.0003
0.08	0.9979	1.0011	0.9989	0.9937	1.0011
0.12	0.9953	1.0024	0.9976	0.9858	1.0024
0.16	0.9916	1.0042	0.9958	0.9747	1.0042
0.20	0.9808	1.0067	0.9934	0.9605	1.0066

(continued)

(continued)

$\alpha l/\pi$	G_1	G_i	G_{vi}	G_v	G_o
0.24	0.9809	1.0096	0.9905	0.9431	1.0096
0.28	0.9739	1.0132	0.9870	0.9226	1.0131
0.32	0.9659	1.0174	0.9830	0.8988	1.0173
0.36	0.9566	1.0222	0.9785	0.8719	1.0220
0.40	0.9462	1.0277	0.9734	0.8418	1.0273
0.44	0.9346	1.0338	0.9677	0.8085	1.0334
0.48	0.9218	1.0407	0.9615	0.7720	1.0401
0.52	0.9077	1.0485	0.9546	0.7323	1.0475
0.56	0.8924	1.0570	0.9472	0.6893	1.0557
0.60	0.8756	1.0665	0.9392	0.6431	1.0647
0.64	0.8574	1.0709	0.9306	0.5937	1.0746
0.68	0.8378	1.0884	0.9213	0.5410	1.0854
0.72	0.8166	1.1010	0.9114	0.4851	1.0972
0.76	0.7938	1.1150	0.9009	0.4258	1.1101
0.80	0.7693	1.1303	0.8896	0.3632	1.1241
0.84	0.7429	1.1471	0.8777	0.2973	1.1394
0.88	0.7147	1.1657	0.8650	0.2281	1.1561
0.92	0.6843	1.1861	0.8516	0.1555	1.1742
0.96	0.6518	1.2087	0.8374	0.0794	1.1941
1.00	0.6108	1.2337	0.8225	0.0000	1.2159
1.04	0.5793	1.2614	0.8067	−0.0829	1.2397
1.08	0.5389	1.2922	0.7900	−0.1693	1.2658
1.12	0.4954	1.3265	0.7725	−0.2592	1.2946
1.16	0.4485	1.3649	0.7540	−0.3527	1.3263
1.20	0.3978	1.4080	0.7345	−0.4499	1.3615
1.24	0.3427	1.4566	0.7140	−0.5506	1.4006
1.28	0.2828	1.5117	0.6924	−0.6551	1.4442
1.32	0.2173	1.5745	0.6697	−0.7634	1.4932
1.36	0.1454	1.6465	0.6458	−0.8755	1.5486
1.40	0.0660	1.7297	0.6206	−0.9915	1.6115
1.44	−0.0223	1.8266	0.5940	−1.1115	1.6835
1.48	−0.1213	1.9405	0.5660	−1.2356	1.7669
1.52	−0.2333	2.0759	0.5364	−1.3638	1.8643
1.56	−0.3616	2.2388	0.5052	−1.4964	1.9795
1.60	−0.5107	2.4379	0.4722	−1.6333	2.1179
1.64	−0.6869	2.6855	0.4372	−1.7749	2.2871
1.68	−0.8999	3.0004	0.4002	−1.9211	2.4987
1.72	−1.1647	3.4122	0.3609	−2.0723	2.7707
1.76	−1.5063	3.9701	0.3192	−2.2286	3.1333
1.80	−1.9695	4.7630	0.2747	−2.3902	3.6407
1.84	−2.6438	5.9692	0.2272	−2.5574	4.4015
1.88	−3.7379	8.0049	0.1764	−2.7305	5.6689
1.92	−5.8769	12.1195	0.1219	−2.9100	8.2013
1.96	−12.1866	24.5630	0.0632	−3.0966	15.8188
2.00	−∞	+∞	0.0	−∞	+∞

For the other second-order correction functions, one has the expressions:

$$G_i(\kappa) = F_i(\kappa) = \frac{1}{2S(\kappa)} \sum_{n=1}^{\infty} \frac{1}{(2n+1)!} \kappa^{n-1}$$

$$G_{vi}(\kappa) = F_{vi}(\kappa) = \frac{1}{6S(\kappa)} \sum_{n=1}^{\infty} \frac{1}{(2n)!} \kappa^{n-1}$$

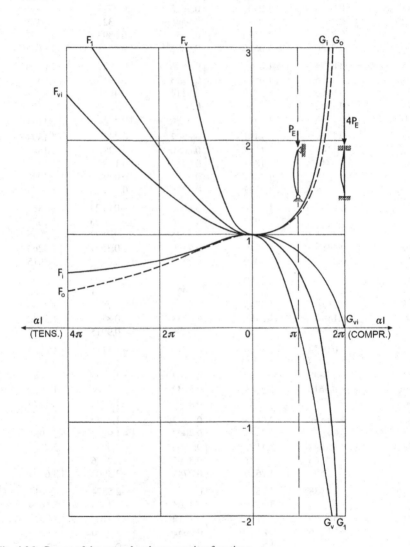

Fig. 4.26 Curves of the second-order correction functions

$$G_v(\kappa) = F_v(\kappa) = \frac{1}{12S(\kappa)} \sum_{n=1}^{\infty} \frac{1}{(2n-1)!} \kappa^{n-1}$$

$$G_o(\kappa) = F_o(\kappa) = 6S(\kappa) / \sum_{n=1}^{\infty} \frac{1}{(2n)!} \kappa^{n-1}$$

In Fig. 4.26, the curves of all these functions are drawn. In the preceding pages, the tables of their numerical values are reported for the ranges of possible use.

4.3.1 Frame Analysis

Let's consider the frame of Fig. 4.27a to be analysed with the second-order theory. Following the criteria of the Displacement Methods and neglecting the axial deformations of the members, this portal can be solved assuming the three geometrical unknowns φ_1, φ_2 and ξ indicated in Fig. 4.27b. Then the following equations are set:

$$\begin{cases} m_1 = 0 \\ m_2 = 0 \\ r = 0 \end{cases}$$

representing the rotation equilibrium of joint 1, the rotation equilibrium of joint 2 and the translation equilibrium of the beam. With the effect superposition, these equations are written as:

$$\begin{cases} (k_{1a} + k_{1b})\varphi_1 + k_{ia}\varphi_2 - k_{vib}\xi + m_{10} = 0 \\ k_{ia}\varphi_1 + (k_{1a} + k_{1c})\varphi_2 - k_{vic}\xi + m_{20} = 0 \\ -k_{vib}\varphi_1 - k_{vic}\varphi_2 + (k_{vb} + k_{vc})\xi + r_0 = 0 \end{cases}$$

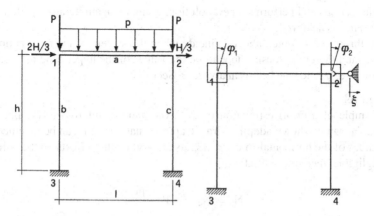

Fig. 4.27 One-storey frame

referring to members with constant section for which the flexural behaviour is represented by the four rotation, translation, direct and indirect stiffness k_1, k_i, k_{vi} and k_v.

Within the second-order theory, these stiffness depend on the axial actions of the members:

$$k = k(\alpha l)$$

with $\alpha^2 = N/EI$. Therefore, the set of equations written above have not constant coefficients. Also their terms due to the loads, related to the flexural behaviour of the members, depend on the axial actions:

$$m_{10} = m_{10}(p; \alpha l)$$

$$m_{20} = m_{20}(p; \alpha l)$$

and so they are again dependent on the unknowns of the problem.

For the members with two full end supports in Sect. 4.2.2 the stiffness and the end moments due to loads have been defined and their expression are now to be used with the proper axial actions for the solution of the present problem.

For the numerical elaboration of the nonlinear set of equations, one can adopt three different levels of refining. At the first level stands the *approximate method* that linearizes the problem adopting, for the evaluation of the coefficients of the equations, an invariable set of axial actions, estimated a priori on the given structural arrangement. At the second level stands the *refined method* that performs a preliminary first-order analysis from which the set of axial actions is deduced, assumed then as invariant for the calculation of the coefficients of the subsequent second-order analysis. At the third level stands the *correct method* that, following the solution of the preceding method, rectifies the coefficients on the basis of the new values of the axial actions and performs e new solution, going on again iteratively till a good convergence is achieved.

The three levels quoted above coincide when the system of the axial actions is isostatic and therefore doesn't depend on the values of the hyperstatic moments of the structure, like for the beams analysed in Sect. 4.2.

First Level
An example of a priori estimation of the axial actions, referred to the portal of Fig. 4.27a, can be shown adopting drastic approximations; if it can be assumed, for low values of the horizontal force H, that its anti-symmetric effects on the columns are negligible, then one can set:

$$N_b = N_c = -P - \frac{pl}{2}$$

If again one assumes that the beam section, dimensioned for the primary bending moment component of the internal action, is little influenced by the small axial component, then one can set:

$$N_a = 0$$

Obviously, these approximations refer to the calculation of the member stiffness. In this case, the beam stiffness would remain unchanged with respect to the first-order values and the columns stiffness rectified with the same correction functions.

An over-estimation of the axial force in the beam can be performed adding, to the action transmitted to distribute the horizontal force on the two columns, the compression coming from the opposite shears induced in the same columns by the fixed end moments of the beam:

$$N_a = -\frac{1}{2}\left(\frac{2H}{3} - \frac{H}{3}\right) - \frac{pl^2}{12}\frac{1.0 + 0.5}{h} = -\frac{H}{6} - \frac{pl^2}{8h}$$

Also the estimation of the axial action in the columns can be improved adding, to the mean value, the anti-symmetric effect of the horizontal force corresponding to the beam shear action, calculated at the upper limit with $V_a = Hh/2l$:

$$N_b = -P - \frac{pl}{2} + \frac{Hh}{2l}$$
$$N_c = -P - \frac{pl}{2} - \frac{Hh}{2l}$$

An indication about the influence of these approximate evaluations on the results of the analysis will be provided with a numerical example reported further on.

Second Level

As already said, following the refined method a preliminary first-order analysis is performed from which a better evaluation of the set of axial actions is deduced, Again with reference to the portal of Fig. 4.27, first a value 1.0 is given to all the involved correction functions here listed:

$$-m_{10} = +m_{20} = \frac{pl^2}{12}G_o(\alpha_a l)$$

$$k_{1a} = \frac{4EI_a}{1}G_1(\alpha_a l) \quad k_{1b} = \frac{4EI_b}{h}G_1(\alpha_b h) \quad k_{1c} = \frac{4EI_c}{h}G_1(\alpha_c h)$$

$$k_{ia} = \frac{2EI_a}{1}G_i(\alpha_a l) \quad k_{ib} = \frac{2EI_b}{h}G_i(\alpha_b h) \quad k_{ic} = \frac{2EI_c}{h}G_i(\alpha_c h)$$

$$k_{via} = \frac{6EI_a}{1^2}G_{vi}(\alpha_a l) \quad k_{vib} = \frac{6EI_b}{h^2}G_{vi}(\alpha_b h) \quad k_{vic} = \frac{6EI_c}{h^2}G_{vi}(\alpha_c h)$$

$$k_{va} = \frac{12EI_a}{1^3}G_v(\alpha_a l) \quad k_{vb} = \frac{12EI_b}{h^3}G_v(\alpha_b h) \quad k_{vc} = \frac{12EI_c}{h^3}G_v(\alpha_c h)$$

Setting for example:

$$l = h \quad I_b = I_c = I \quad I_a = 4\,I$$

one has the following set of equations (with $\bar{\xi} = \xi/h$):

$$
\begin{cases}
+\dfrac{2EI}{l}(8+2)\varphi_1 + \dfrac{2EI}{l}4\varphi_2 - \dfrac{2EI}{l}3\bar{\xi} = +\dfrac{pl^2}{12} \\[4mm]
+\dfrac{2EI}{l}4\varphi_1 + \dfrac{2EI}{l}(8+2)\varphi_2 - \dfrac{2EI}{l}3\bar{\xi} = -\dfrac{pl^2}{12} \\[4mm]
-\dfrac{2EI}{l}3\varphi_1 - \dfrac{2EI}{l}3\varphi_2 + \dfrac{2EI}{l}(6+6)\bar{\xi} = +H
\end{cases}
$$

that becomes, after the pertinent simplifications:

$$
\begin{cases}
+10\varphi_1 + 4\varphi_2 - 3\bar{\xi} = +\dfrac{pl^3}{24EI} \\[4mm]
+4\varphi_1 + 10\varphi_2 - 3\bar{\xi} = -\dfrac{pl^3}{24EI} \\[4mm]
-3\varphi_1 - 3\varphi_2 + 12\bar{\xi} = +\dfrac{pl^3}{24EI}\dfrac{12H}{pl}
\end{cases}
$$

Setting for the case of concern:

$$H = \frac{1}{6}\frac{pl}{2} \quad \left(\frac{12H}{pl} = 1\right)$$

the solution "1" is:

$$\varphi_1 = +0.09333\overline{M} \cdot 1/EI$$

$$\varphi_2 = -0.07333\overline{M} \cdot 1/EI$$

$$\bar{\xi} = +0.04667\overline{M} \cdot 1/EI$$

where $\overline{M} = pl^2/12$. So one has the following moments:

$$M_{a1} = -\overline{M} + \frac{4EI_a}{l}\varphi_1 + \frac{2EI_a}{l}\varphi_2 = -0.0933\overline{M}$$

$$M_{a2} = -\overline{M} - \frac{2EI_a}{l}\varphi_1 - \frac{4EI_a}{l}\varphi_2 = -0.5733\overline{M}$$

$$M_{ao} = +\frac{3}{2}\overline{M} + \frac{M_{a1} + M_{a2}}{2} = +1.1667\overline{M}$$

$$M_{b1} = -\frac{4EI_b}{h}\varphi_1 + \frac{6EI_b}{h^2}\xi = -0.0933\overline{M}$$

$$M_{b3} = +\frac{2EI_b}{h}\varphi_1 - \frac{6EI_b}{h^2}\xi = -0.0933\overline{M}$$

$$M_{c2} = +\frac{4EI_c}{h}\varphi_2 - \frac{6EI_c}{h^2}\xi = -0.5733\overline{M}$$

$$M_{c4} = -\frac{2EI_c}{h}\varphi_2 + \frac{6EI_c}{h^2}\xi = +0.4267\overline{M}$$

and the following shear forces:

$$V_{a1} = +\overline{V} - \frac{M_{a1} - M_{a2}}{l} = +0.9200\overline{V}$$

$$V_{a2} = -\overline{V} - \frac{M_{a1} - M_{a2}}{l} = -1.0800\overline{V}$$

$$V_b = -\frac{M_{b1} - M_{b3}}{h} = 0$$

$$V_c = -\frac{M_{c2} - M_{c4}}{h} = +0.1667\overline{V}$$

So eventually, setting $\overline{N} = \overline{V} = pl/2$ one obtains (with $P = 0 = $ the axial actions):

$$N_b = -V_{a1} = -0.9200\overline{N}$$

$$N_c = -V_{a2} = -1.0800\overline{N}$$

$$N_a = -\frac{2}{3}H + V_b = \frac{1}{3}H - V_c = -0.1111\overline{N}$$

With the estimations proposed before for the approximate method, the following axial actions would be assumed:

$$N_b = -\frac{pl}{2} = -1.0000\overline{N}$$

$$N_c = -\frac{pl}{2} = -1.0000\overline{N}$$

$$N_a = 0$$

possibly improved in:

$$N_b = -\frac{pl}{2} + \frac{Hh}{2l} = -0.9167\overline{N}$$

$$N_c = -\frac{pl}{2} + \frac{Hh}{2l} = -1.0833\overline{N}$$

$$N_a = -\frac{H}{6} - \frac{pl^2}{8h} = -0.2778\overline{N}$$

The influence of these differences of evaluation of the axial forces on the coefficients of the solving equations and consequently on the hyperstatics depends on the intensity of the same axial forces as a ratio to the member stiffness parameter EI/l^2 that is to their Euler's critical load $\pi^2 EI/l^2$.

Let's consider for example the three intensities of axial force:

$$v = \frac{N}{N_E} = \frac{pl/2}{\pi^2 EI/h^2} = 0.2 - 0.4 - 0.6$$

acting on the columns of the portal under examination. For the axial actions before calculated with the first-order analysis "1" (refined method), with the a priori estimation "e" (approximate method) and with an improved estimation "s" of the same approximate method, one has the correction factors of the following tables. They are referred to the corresponding terms of the equations written for $v = 0$ (first-order analysis—coefficients and known terms) and to the related solution. The details of the elaborations have been omitted. They start from the calculation of the functions "G" of the members and proceed with the elaborations of the "new" equations, up to the evaluation of the unknowns, providing the refined and approximate solution of the second-order analysis "2".

– **Analysis "2" for $v = 0.2$**

From "1"	0.986	1.001	0.993	1.001	1.051
	1.001	0.984	0.964	1.001	0.999
	0.993	0.964	0.802	1.000	1.276
From "e"	0.986	1.000	0.967	1.000	1.047
	1.000	0.986	0.967	1.000	0.993
	0.967	0.967	0.802	1.000	1.273
From "s"	0.984	1.002	0.969	1.002	1.056
	1.002	0.982	0.964	1.002	1.006
	0.969	0.964	0.802	1.000	1.277

– **Analysis "2" for $v = 0.4$**

From "1"	0.972	1.002	0.938	1.002	1.124
	1.002	0.967	0.927	1.002	0.977
	0.938	0.927	0.603	1.000	1.765
From "e"	0.972	1.000	0.932	1.000	1.115
	1.000	0.972	0.932	1.000	0.964
	0.932	0.932	0.603	1.000	1.757
From "s"	0.967	1.005	0.938	1.005	1.137
	1.005	0.962	0.926	1.005	0.991
	0.938	0.926	0.603	1.000	1.766

– Analysis "2" for $\nu = 0.6$

From "1"	0.956	1.003	0.905	1.003	1.260
	1.003	0.949	0.888	1.003	0.887
	0.905	0.888	0.403	1.000	2.861
From "e"	0.957	1.000	0.898	1.000	1.243
	1.000	0.957	0.898	1.000	0.866
	0.898	0.898	0.403	1.000	2.839
From "s"	0.950	1.007	0.906	1.007	1.281
	1.007	0.942	0.888	1.007	0.910
	0.906	0.888	0.403	1.000	2.865

One can notice that the differences are small for the lower intensities, while the higher axial actions, those that can arise in frames with very slender columns, analysed with the increased levels of loads for the resistance verifications, lead to big corrections of the results. Less influent appears the approximation level of the a priori estimation of axial actions.

Similar tables are given below in terms of moments at the frame joints, having reproduced in the first line those coming from the first-order solution.

– Solution "2" for $\nu = 0.2$

	M_{a1}	M_{ao}	M_{a2}	M_{b3}	M_{c4}
	−0.0933	+1.1667	−0.5733	−0.0933	+0.4267
From "1"	−0.0215	+1.1876	−0.6161	−0.144	+0.4966
From "e"	−0.0196	+1.182	−0.6163	−0.1425	+0.4955
From "s"	−0.0235	+1.1954	−0.6179	−0.143	+0.4977

– Solution "2" for $\nu = 0.4$

	M_{a1}	M_{ao}	M_{a2}	M_{b3}	M_{c4}
	−0.0933	+1.1667	−0.5733	−0.0933	+0.4267
From "1"	−0.0970	+1.2111	−0.7011	−0.2392	+0.6128
From "e"	+0.1001	+1.1991	−0.7019	−0.2349	+0.6105
From "s"	+0.0932	+1.2275	−0.7048	−0.2372	+0.6154

– Solution "2" for ν = 0.6

	M_{a1}	M_{ao}	M_{a2}	M_{b3}	M_{c4}
	−0.0933	+1.1667	−0.5733	−0.0933	+0.4267
From "1"	+0.3471	+1.2383	−0.9109	−0.4644	+0.8588
From "e"	+0.3486	+1.2181	−0.9124	−0.453	+0.8553
From "s"	+0.3419	+1.2643	−0.9166	−0.4616	+0.8634

Third Level

Further to the solution "2" elaborated with the values "1" of the axial actions, following the correct method the procedure can be resumed with the new values of the axial actions with which the corresponding functions "G" are calculated and a subsequent elaboration of the equations is performed. So a new solution with the related new axial actions is obtained to go on iteratively until the differences between the new and the preceding solution become negligible.

The procedure is shown in the following with reference to the same portal of Fig. 4.27a and to the intensity ν = 0.6 of the axial action (mean on the columns). For this high intensity, there are relevant second-order contributions. In the following table, for five subsequent solutions, the values of the principal output data are reported: the geometric unknowns, the bending moments at the joints of the portal and at mid-span of the beam, and the axial actions on the members. From these last values starts the analysis synthetized by the results of the subsequent column.

	"0"	"1"	"2"	"3"	"4"	"5"
φ_1		+0.09333	+0.11764	+0.11680	+0.11674	+0.11674
φ_2		−0.07333	−0.06508	−0.06466	−0.06461	−0.06461
$\overline{\xi}$		+0.04667	+0.13350	+0.13490	+0.13488	+0.13488
M_{a1}		−0.0933	+0.3471	+0.3547	+0.3548	+0.3548
M_{ao}		+1.1667	+1.2343	+1.2243	+1.2234	+1.2234
M_{a2}		−0.5733	−0.9109	−0.8974	−0.8973	−0.8973
M_{b3}		−0.0933	−0.4644	−0.4891	−0.4891	−0.4891
M_{c4}		+0.4267	+0.8588	+0.8564	+0.8563	+0.8563
N_a	0.0000	−0.1111	+0.0241	+0.0295	+0.0295	
N_b	0.0000	−0.9200	−0.7903	−0.7913	−0.7913	
N_c	0.0000	−1.0800	−1.2097	−1.2087	−1.2087	

One can notice the good convergence or the procedure that leads to relative errors lower than 10^{-4} after only four iterations.

In conclusion of this discourse, the diagrams of the internal forces are reported, drawing the relative graphs in Fig. 4.28 where, only for the intensity ν = 0.6, the "correct" second-order solution is indicated with a continuous line superimposed to

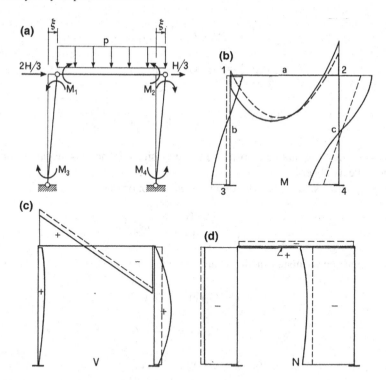

Fig. 4.28 Second-order diagrams of internal forces

the first-order dashed one. For the analysis of the single members, after the calculation of the hyperstatics, the algorithms for the superposition of effects presented at Sect. 4.1.2 (see Fig. 4.10) have been applied.

4.3.2 Linearized Theory

The second-order linearized theory replaces the expressions of the correction functions with their Mc Laurin's series development on the quantity

$$\kappa = (\alpha l)^2 = N l^2 / E I$$

truncating the same development to the first two terms. So for the stiffness of the doubly fixed end member one has the following approximate correction functions:

$$F_1(\kappa) = G_1(\kappa) = 1 + \frac{1}{30}\kappa$$

$$F_i(\kappa) = G_i(\kappa) = 1 - \frac{1}{60}\kappa$$

$$F_{vi}(\kappa) = G_{vi}(\kappa) = 1 + \frac{1}{60}\kappa$$

$$F_v(\kappa) = G_v(\kappa) = 1 + \frac{1}{10}\kappa$$

again assuming the quantity κ positive for the tensile axial actions, negative for the compressive axial actions.

Setting, with a small safe side round off, $\pi^2 \approx 10$, one has:

$$v = \frac{N}{N_E} = \frac{N}{\pi^2 EI/l^2} = \kappa/10$$

and the same approximate functions can be written as:

$$F_1(v) = G_1(v) = 1 + v/3$$

$$F_i(v) = G_i(v) = 1 - v/6$$

$$F_{vi}(v) = G_{vi}(v) = 1 + v/6$$

$$F_v(v) = G_v(v) = 1 + v$$

assuming the dimensionless axial action v, ratio to Euler's critical load, positive if in tension, negative if in compression.

The approximation degree of these linearized functions can be deduced from the diagrams of Fig. 4.29 where, for the four flexural stiffness of the member, to the correct curve the corresponding straight (dashed) line is indicated.

One can notice that, except for small values of the quantity v, the rotation stiffnesses $F_1 - G_1$ and $F_i - G_i$ are badly approximated by the corresponding straight lines. In the range of the compressions, the differences between the curves and the straight lines are such as to remove any validity to the linear approximation when the value $v = -1$ is overcome.

On the contrary, the translation stiffnesses $F_{vi} - G_{vi}$ and $F_v - G_v$ are better approximated by the corresponding straight lines. Mainly, the direct translation stiffness $F_v - G_v$ remains well approximated, also in the compression range, over all the interval $0 > v - 4$ of possible use.

From these outcomes, it follows that in the frame analysis the linear approximation gives good results if the translation behaviour evidenced by large joint displacements is prevailing. If instead the rotation behaviour remains prevalent, as happens for the non-sway frames, then the second-order linearized theory can lead to big errors. For

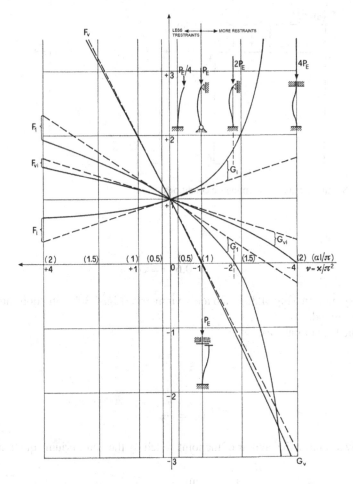

Fig. 4.29 Correction functions curves

low values of the axial actions v the linear approximation is good, but the analysis moves with small second-order contributions and so the discourse looses interest.

The principal usefulness of the binomial linear expressions is in the instability calculations that will be presented in the following section. These calculations refer to limit states with high axial compressions. So the problem of the good approximation will have to be resumed and deepened. For the moment, only a simple numerical application is presented in order to give a qualitative indication of the possible differences of the results.

So let's consider the same portal analysed at the preceding section (see Fig. 4.27). Assuming an approximate and invariant set of axial actions with:

$$v_a = 0 \; (=\text{const.})$$

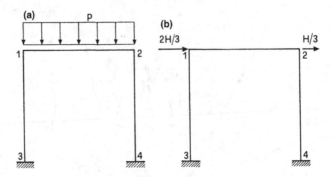

Fig. 4.30 Symmetric and anti-symmetric conditions

$$v_b = v_c = 0.6 \quad (= \text{const.})$$

the two symmetric (Fig. 4.30a) and anti-symmetric (Fig. 4.30b) situations are separately examined.

For the former one has:

$$\xi = 0$$

$$\varphi_1 = -\varphi_2$$

with only a rotation behaviour of the joints ruled by the equilibrium equation:

$$m_1 = 0$$

So with reference to the same data assumed at Sect. 4.2.1, one has (with $G_{1a} = G_{ia} = G_{oa} = 1$):

$$\left(\frac{16EI}{1} + \frac{4EI}{1}G_{1b}\right)\varphi_1 - \frac{8EI}{1}\varphi_1 - \frac{pl^2}{12} = 0$$

that leads to the following solutions:

– **First order** ($\overline{M} = pl^2/12$):

$$G_{1b} = 1$$

$$\frac{12EI}{1}\varphi_1 - \frac{pl^2}{12} = 0$$

$$\varphi_1 = \frac{pl^3}{12^2 EI} = \overline{\varphi}_1$$

$$M_1 = -\frac{4}{EI}l\varphi_1 = -\frac{1}{3}\overline{M} = \overline{M}_1$$

$$M_3 = +\frac{2}{EI}l\varphi_1 = +\frac{1}{6}\overline{M} = \overline{M}_3$$

– Second order linearized ($\nu_b = -0.6$):

$$G_{1b} = 1 + \nu_b/3 = 0.8000$$

$$\frac{110.20EI}{1}\varphi_1 - \frac{pl^2}{12} = 0$$

$$\varphi_1 = 10.0714\overline{\varphi}_1$$

$$M_1 = -\frac{4EI}{1}G_{1b}\varphi_1 = 0.8571\overline{M}_1$$

$$G_{ib} = 1 - \nu_b/6 = 1.1000$$

$$M_3 = +\frac{2EI}{1}G_{ib}\varphi_1 = 1.1785\overline{M}_3$$

– Second-order correct ($\kappa_b = \pi^2 \nu_b$):

$$G_{1b} = G_1(\kappa_b) = 0.7851$$

$$\frac{11.14EI}{1}\varphi_1 - \frac{pl^2}{12} = 0$$

$$\varphi_1 = 1.0772\overline{\varphi}_1$$

$$M_1 = -\frac{4EI}{1}G_{1b}\varphi_1 = 0.8457\overline{M}_1$$

$$G_{ib} = G_i(\kappa_b) = 1.1204$$

$$M_3 = +\frac{2EI}{1}G_{ib}\varphi_1 = 1.2069\overline{M}_3$$

For the latter situation one has:

$$\xi \neq 0$$

$$\varphi_1 = \varphi_2$$

with prevalent translation behaviour of the joints ruled by the equilibrium equations:

$$\begin{cases} m_1 = 0 \\ r = 0 \end{cases}$$

Again under the same axial actions one has (with $\bar{\xi} = \xi/l$ and with $G_{vic} = G_{vib}$, $G_{vc} = G_{vb}$):

$$\begin{cases} \left(\dfrac{16EI}{1} + \dfrac{4EI}{1} G_{1b} \right) \varphi_1 + \dfrac{8EI}{1} \varphi_1 - \dfrac{6EI}{1} G_{vib}\bar{\xi} = 0 \\[3mm] -\dfrac{6EI}{1^2} G_{vib}\varphi_1 - \dfrac{6EI}{1^2} G_{vib}\varphi_1 + \dfrac{24EI}{1^2} G_{vb}\bar{\xi} - H = 0 \end{cases}$$

that lead to the following solutions:

– **First-order** ($\overline{M} = Hh/4$):

$$G_{1b} = G_{vib} = G_{vb} = 1$$

$$\begin{cases} \dfrac{28EI}{1} \varphi_1 - \dfrac{6EI}{1}\bar{\xi} = 0 \\[3mm] -\dfrac{12EI}{1} \varphi_1 + \dfrac{24EI}{1}\bar{\xi} = Hl \end{cases}$$

$$\varphi_1 = \dfrac{Hl^2}{100EI} = \dfrac{1}{25}\dfrac{1}{EI}\overline{M} = \overline{\varphi}_1$$

$$\bar{\xi} = \dfrac{7Hl^2}{150EI} = \dfrac{14}{75}\dfrac{1}{EI}\overline{M} = \bar{\xi}_o$$

$$M_1 = -\dfrac{4EI}{1} \varphi_1 + \dfrac{6EI}{1}\bar{\xi} = +\dfrac{24}{25}\overline{M} = \overline{M}_1$$

$$M_3 = +\dfrac{2EI}{1} \varphi_1 - \dfrac{6EI}{1}\bar{\xi} = -\dfrac{26}{25}\overline{M} = \overline{M}_3$$

– **Second-order linearized** ($v_b = -0.6$):

$$G_{1b} = 1 + v_b/3 = 0.800$$

$$G_{v1b} = 1 + v_b/6 = 0.900$$

$$G_{vb} = 1 + v_b = 0.400$$

$$\begin{cases} +\dfrac{27.20EI}{1}\varphi_1 - \dfrac{5.40EI}{1}\bar{\xi} = 0 \\[2mm] -\dfrac{10.80EI}{1}\varphi_1 + \dfrac{9.60EI}{1}\bar{\xi} = Hl \end{cases}$$

$$\varphi_1 = 2.6627\bar{\varphi}_1$$

$$\bar{\xi} = 2.8740\bar{\xi}_o$$

$$M_1 = -\frac{4EI}{1}G_{1b}\varphi_1 + \frac{6EI}{1}G_{vib}\bar{\xi} = 2.6628\overline{M}_1$$

$$G_{ib} = 1 - v_b/6 = 1.1000$$

$$M_3 = +\frac{2EI}{1}G_{ib}\varphi_1 - \frac{6EI}{1}G_{vib}\bar{\xi} = 2.5604\overline{M}_3$$

– **Second-order correct** ($\kappa_b = \pi^2 v_b$):

$$G_{1b} = G_1(\kappa_b) = 0.7851$$

$$G_{vib} = G_{vi}(\kappa_b) = 0.8968$$

$$G_{vb} = G_v(\kappa_b) = 0.4034$$

$$\begin{cases} \dfrac{27.14EI}{1}\varphi_1 - \dfrac{5.38EI}{1}\bar{\xi} = 0 \\[2mm] -\dfrac{10.76EI}{1}\varphi_1 + \dfrac{9.68EI}{1}\bar{\xi} = Hl \end{cases}$$

$$\varphi_1 = 2.6266\bar{\varphi}_1$$

$$\bar{\xi} = 2.8393\bar{\xi}_o$$

$$M_1 = -\frac{4EI}{1}G_{1b}\varphi_1 + \frac{6EI}{1}G_{vib}\bar{\xi} = 2.6271\overline{M}_1$$

$$G_{ib} = G_i(\kappa_b) = 1.1204$$

$$M_3 = +\frac{2EI}{1}G_{ib}\varphi_1 - \frac{6EI}{1}G_{vib}\bar{\xi} = 2.5159\overline{M}_3$$

From the numerical examples presented above, the following considerations can be deduced. First, being in a level of axial actions still far from the limit $v = -1$, also for the rotation behaviour the second-order linearized analysis gives good results, with errors of few per cent. Second, the different influence of the second-order contributions is evident, being sensible also for the rotation behaviour (up to 20% in terms of moment), but reaching much larger values for the translation behaviour, even greater than the first-order ones (up to 160% in terms of moments):

This latter unfavourable effect is due to the action of the vertical gravity loads on the deformed structure with sensible horizontal translation of the beam, as numerically can be shown by its equilibrium condition:

$$2\frac{M_{b1} - M_{b3}}{1} + 2N_b\frac{\xi}{1} = H$$

With $N_b = -0.6N_E$ and with the values of the last solution (second-order correct), this condition becomes:

$$2\frac{2.5220 + 2.6165}{1}\overline{M} + 2\frac{N_b}{EI/1}0.5300\overline{M} = H$$

where $\overline{M} = Hl/4$, leading to:

$$2.5693 - 1.5693 = 1.0000$$

That is, the moment M_{b1}, M_{b3} on the columns, further to the equilibrium versus the horizontal force H, shall provide support to the deviatory action of the vertical loads acting on the displaced configuration of the structure (see Fig. 4.28a). To the action of the applied horizontal force H is devoted a share

$$\frac{1.0000}{2.5693}100 = 39\%$$

of the column resistance; to the deviatory action of the vertical loads is devoted a share:

$$\frac{1.5693}{2.5693}100 = 61\%$$

These high deviatory actions make the sway multi-storey frames much more sensible to second-order effects and the related possible lateral instability failures than the non-sway frames provided with bracing systems.

Eventually, an alternative form of the second-order linearized expressions is presented, where the basic first-order contribution of the stiffness is separated from the incremental term due to the axial action, with expressions such as:

$$k = k_o + \Delta k$$

So for the four stiffnesses of the beam with double fixed ends, the following expressions can be used:

$$k_1 = \frac{4EI}{1}G_1(\kappa) = \frac{4EI}{1} + \frac{21}{15}N$$

$$k_i = \frac{2EI}{1}G_i(\kappa) = \frac{2EI}{1} - \frac{1}{30}N$$

$$k_{vi} = \frac{6EI}{1^2}G_{vi}(\kappa) = \frac{6EI}{1^2} + \frac{1}{10}N$$

$$k_v = \frac{12EI}{1^3}G_v(\kappa) = \frac{12EI}{1^3} + \frac{6}{51}N$$

with N positive if in tension.

4.3.3 Elastic Instability of Frames

The same algorithm presented in the preceding sections for the second-order analysis of frames is used also for their instability calculations. In the former case, for a given load condition, the structure has been solved up to the definition of the internal forces. Now instead one has to calculate the load intensity that leads the structure to the instability limit. In the former case, a set of equilibrium equations had to be solved having as unknowns the joint displacements. In the latter case, a proper equation has to be solved, deduced from the same set of equations as "instability condition", where the unknown is a parameter γ related to the intensity of the acting loads.

Let's consider the frame of Fig. 4.27: its behaviour is governed by the equilibrium equations written at the beginning of Sect. 4.3.1. The condition to have effects φ_1, φ_2, ξ infinite under the action of finite loads is simply set equalizing the determinant of the coefficient matrix to zero:

$$\text{det.} = 0$$

Assuming as unknowns the quantities κ_j of the stiffness correction functions, related to the axial actions of the frame members, this condition becomes the solving equation of the problem.

A solution of this equation can be calculated reducing the number of unknowns to one by fixing the ratio among the axial actions of the different members. That is, starting from a given set N_1, N_2, ..., N_j, ..., N_m of axial actions, for all these an increase proportional to one only parameter γ is assumed with:

$$\kappa_j = \gamma\frac{N_j l_j^2}{EI_j}$$

where the *critical factor* γ remains the only unknown of the problem.

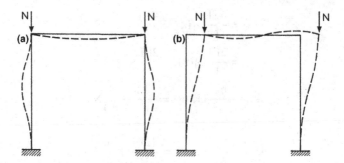

Fig. 4.31 Instability failure modes symmetric (**a**) anti-symmetric (**b**)

An equivalent unknown reduction can be set assuming as reference parameter the axial action N of a given member, e.g.: the most loaded one ($N = N_i$,) and expressing the others as ratio to it:

$$\kappa_j = N \frac{n_j l_j^2}{EI_j}$$

where the quantities $n_j = N_j/N$ remain invariable at the load increase.

Some applications of such calculation are shown first with reference to the simple cases of symmetric and anti-symmetric behaviour presented at Sect. 4.3.2. If a symmetric failure mode is assumed like the one indicated in Fig. 4.31a, and assuming again

$$v_a = 0 \quad (=\text{const.})$$

$$v_b = v_c = v \quad (\text{variable})$$

then the only equilibrium equation:

$$\left(\frac{4EI}{l} + \frac{4EI}{l} G_1 \right) \varphi_1 = \frac{pl^2}{12}$$

leads to the instability condition:

$$\frac{4EI}{l} + \frac{4EI}{l} G_1(\kappa) = 0$$

with $\kappa = \pi^2 v$.

So one has, after the pertinent simplifications, the transcendental equation:

$$G_1(\kappa) + 2 = 0$$

of which, being the possible deformed shape of the columns included between that of the double fixed ends ($\overline{\kappa} = -4\pi^2$) and that of a fixed and a hinged end ($\overline{\kappa} = -2\pi^2$), the root between these two limits has to be found:

$$-2\pi^2 > \kappa > -4\pi^2$$

Calculated numerically by trial the solution is

$$\overline{\kappa} = -32.1$$

that leads to the critical load:

$$-v'_E = -\overline{\kappa}/\pi^2 = 3.25$$

With $\overline{\alpha}1 = \sqrt{|\overline{\kappa}|} = 5.67$ the supports factor is:

$$\beta = \pi/\overline{\alpha}1 = 0.55$$

close to that of the double fixed ends column.

If expressed with the linearized theory, the same instability condition is approximated to the algebraic equation:

$$\left(1 + \frac{1}{30}\kappa\right) + 2 = 0$$

that gives a root:

$$\kappa = -90$$

largely mistaken. Being much beyond the limit $v = -1$ (see Fig. 4.29), the linear approximation for the correction function G_1 is not valid. It would give the impossible solution of a critical load greater than that of the doubly fixed ends column (with $-v'_E = 9.12$ and $\beta = 0.33$).

If, for the same frame, an anti-symmetric failure mode is assumed like the one indicated in Fig. 4.31b (again with $v_a = 0$ and $v_b = v_c = v$), then one has the two equilibrium equations:

$$\begin{cases} \left(\dfrac{24EI}{1} + \dfrac{4EI}{1}G_1\right)\varphi_1 - \dfrac{6EI}{1}G_{vi}\overline{\xi} = 0 \\[4mm] -\dfrac{12EI}{1^2}G_{vi}\varphi_1 + \dfrac{24EI}{1^2}G_v\overline{\xi} = H \end{cases}$$

that can be simplified into:

$$\begin{cases} 2(6 + G_1)\varphi_1 - 3G_{vi}\bar{\xi} = 0 \\[2mm] -G_{vi}\varphi_1 + 2G_v\bar{\xi} = \dfrac{Hl^2}{12EI} \end{cases}$$

Putting the determinant of the coefficient matrix equal to zero, one obtains the transcendental equation:

$$4[6 + G_1(\kappa)]G_v(\kappa) - 3G_{vi}^2 = 0$$

of which, being the possible deformed shape of the columns included between that of the member with a full and a rotation restraints ($\bar{\kappa} = -1\pi^2$) and that of a cantilever member ($\bar{\kappa} = -0.25\pi^2$), the root between these two limits has to be found:

$$-0.25\pi^2 > \kappa > -\pi^2$$

Calculated numerically by trial the solution is

$$\bar{\kappa} = -9.096$$

that leads to the critical load:

$$-v'_E = -\bar{\kappa}/\pi^2 = 0.92$$

With $\bar{\alpha}1 = \sqrt{|\bar{\kappa}|} = 3.02$ the supports factor is:

$$\beta = \pi/\bar{\alpha}1 = 1.04$$

close to that of the column with a full and a rotation restraints.

   If expressed with the linearized theory, the same instability condition is approximated to the algebraic equation:

$$4\left(7 + \frac{1}{30}\kappa\right)\left(1 + \frac{1}{10}\kappa\right) - 3\left(1 + \frac{1}{60}\kappa\right)^2 = 0$$

that can be simplified to:

$$45\kappa^2 + 10200\kappa + 90000 = 0$$

and gives the first root:

$$\kappa = \frac{-5100 + \sqrt{5100^2 - 45 \times 90{,}000}}{45} = -9.20$$

with the critical load:

$$-v'_E = -\bar{\kappa}/\pi^2 = 0.93$$

Being on the smaller values of the load corresponding to less restrained support arrangements (see Fig. 4.28), but above all dealing with a prevalent translation behaviour, the very good precision of the linearized theory can be noticed with results practically coinciding with the correct ones.

The analysis repeated for both the cases symmetric and anti-symmetric served to present the calculation procedure and clarify the ranges of applicability of the linear approximations. For the application ends, one could foresee a priori what was the first failure mode of the portal under examination, the one associated with the lowest value of N (assumed as *critical load*). In fact looking at Fig. 4.30b and reminding the elementary cases of column instability treated at Sect. 4.2, the symmetric failure mode showed for the columns a buckling length included between that of the double fixed ends and that of a fixed and a hinged end with $0.51 < l_o < 0.71$, the anti-symmetric failure mode showed for the columns a buckling length included between that of a full and a rotation restraints and that of a cantilever arrangement with $1 < l_o < 2.0l$. So to this latter mode, related to higher column slenderness, one could limit the analysis, expecting certainly a lower value of the critical load.

For more complex structures in general, it is not possible to simplify the analysis using conditions of symmetry and anti-symmetry. So one has to proceed with the whole set of equations putting the determinant of its coefficient matrix equal to zero. For example, with reference to the same portal of Fig. 4.27 and with the same assumption of $v_a = 0$ and $v_b = v_c = v$, the homogeneous form of the equilibrium equation set (having taken the known terms off) is:

$$\begin{cases} \dfrac{2EI}{1}(8 + 2G_1)\varphi_1 + \dfrac{2EI}{1}4\varphi_2 - \dfrac{2EI}{1}3G_{vi}\bar{\xi} = 0 \\[3mm] \dfrac{2EI}{1}4\varphi_1 + \dfrac{2EI}{1}(8 + 2G_1)\varphi_2 - \dfrac{2EI}{1}3G_{vi}\bar{\xi} = 0 \\[3mm] -\dfrac{2EI}{1}3G_{vi}\varphi_1 - \dfrac{2EI}{1}3G_{vi}\varphi_2 + \dfrac{2EI}{1}12G_v\bar{\xi} = 0 \end{cases}$$

Deleting the constant $3EI/l$, the instability condition is set with the equation:

$$\det \begin{vmatrix} 8 + 2G_1(\kappa) & 4 & -3G_{vi}(\kappa) \\ 4 & 8 + 2G_1(\kappa) & -3G_{vi}(\kappa) \\ -3G_{vi}(\kappa) & -3G_{vi}(\kappa) & 12G_v(\kappa) \end{vmatrix} = 0$$

For the solution of this equation, one can repeat the evaluation of the determinant with values of the quantity κ chosen by trial, possibly adopting proper procedures to orient the trials and with the possibility to determine a priori the range in which the

Fig. 4.32 Curves of the
determinant values

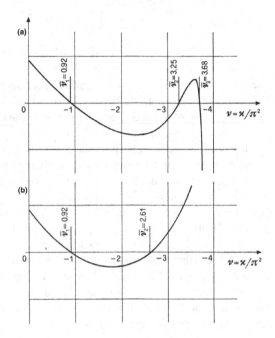

solution $\bar{\kappa}$ is to be searched. The solution will be represented by the value of the last trial for which with good precision the equality det. $= 0$ is found.

There are complex problems of numerical calculation related to this procedure, mainly when being applied to big structures, an optimization for time/precision of the automated elaboration is required. Belonging to the competences of another discipline, these problems are not treated in this text.

For the case under examination, for which a matrix of small dimensions is concerned, the whole curve of the determinant value is provided in the range $0 \geq \kappa > -4\pi^2$ (see Fig. 4.32a). One can notice how the solutions:

$$\bar{\kappa}_1 = -9.096 \; (\bar{v}_1 = 0.92)$$
$$\bar{\kappa}_2 = -32.11 \; (\bar{v}_2 = 3.25)$$

are represented, already previously found by means of simplified algorithms. From the general one, within the given range, all the roots there included come out. These roots correspond to the different possible instability modes and are indicated as *eigenvalues*.

To find the deformed shape of the structure for any eigenvalue, in the original set of homogeneous equations an arbitrary value is given to one of the joint unknowns (e.g.: an unity value) and the remaining ones are consequently calculated. So for example, if for the first eigenvalue one sets $\bar{\xi} = 1$, by consequence one has a set reduced to the first two equations:

$$\begin{cases} [8 + 2G_1(\overline{\kappa}_1)]\varphi_1 + 4\varphi_2 = 3G_{vi}(\overline{\kappa}_1) \\ 4\varphi_1 + [8 + 2G_1(\overline{\kappa}_1)]\varphi_2 = 3G_{vi}(\overline{\kappa}_1) \end{cases}$$

that, with:

$$G_1(\overline{\kappa}_1) = 0.652$$

$$G_{vi}(\overline{\kappa}_1) = 0.837$$

becomes:

$$\begin{cases} 9.304\varphi_1 + 4.000\varphi_2 = 2.512 \\ 4.000\varphi_1 + 9.304\varphi_2 = 2.512 \end{cases}$$

and leads to:

$$\varphi_1 = \varphi_2 = 0.189$$

Together with the unity value of the third unknown, these latters forms the first *eigenvector*:

$$\begin{vmatrix} +0.189 \\ +0.189 \\ +1.000 \end{vmatrix}$$

from which, being $\varphi_1 = \varphi_2 > 0$ and $\xi > 0$, one deduces that an anti-symmetric deformed shape is represented such as that of Fig. 4.31b.

Similarly for the second eigenvalue, one sets $\varphi_1 = 1$ obtaining a set reduced to the last two equations:

$$\begin{cases} [8 + 2G_1(\overline{\kappa}_2)]\varphi_2 - 3G_{vi}(\overline{\kappa}_2)\overline{\xi} = -4 \\ -3G_{vi}(\overline{\kappa}_2)\varphi_2 + 12G_v(\overline{\kappa}_2)\overline{\xi} = +3G_{vi}(\overline{\kappa}_2) \end{cases}$$

that, with:

$$G_1(\overline{\kappa}_2) = -2.00$$

$$G_{vi}(\overline{\kappa}_2) = +0.27$$

$$G_v(\overline{\kappa}_2) = -2.41$$

becomes:

$$\begin{cases} 4.00\varphi_2 - 0.81\bar{\xi} = -4.00 \\ 0.81\varphi_2 + 28.90\bar{\xi} = +0.81 \end{cases}$$

and leads to:

$$\varphi_2 = -1.000 \quad (= -\varphi_1)$$

$$\bar{\xi} = 0$$

So the second eigenvector is:

$$\begin{vmatrix} +1.000 \\ -1.000 \\ 0.000 \end{vmatrix}$$

and shows a symmetric deformed shape such as that of Fig. 4.31a.

The examination of the portal is concluded showing the influence on the critical load of a dissymmetry of the axial forces, like the one coming from the horizontal action H that makes a column more compressed than the other. Let's consider for example the improved estimation of the approximate method made at Sect. 4.3.1:

$$v_a = 0.2778v = \kappa_a/\pi^2$$

$$v_b = 0.9167v = \kappa_b/\pi^2$$

$$v_c = 1.0833v = \kappa_c/\pi^2$$

In this case, the instability condition is set with the equation:

$$\det \begin{vmatrix} 8G_1(\kappa_a) + 2G_1(\kappa_b) & 4G_i(\kappa_a) & -3G_{vi}(\kappa_b) \\ 4G_i(\kappa_a) & 8G_1(\kappa_a) + 2G_1(\kappa_c) & -3G_{vi}(\kappa_c) \\ -3G_{vi}(\kappa_b) & -3G_{vi}(\kappa_c) & 6G_v(\kappa_b) + 6G_v(\kappa_c) \end{vmatrix} = 0$$

from which one can calculate for the unknown $\kappa = \pi^2 v$ the eigenvalues, as indicated in Fig. 4.32b by means of the diagram of the determinant drawn for the whole range $0 \geq \kappa > -4\pi^2$. The two eigenvalues obtained from the diagram are:

$$\bar{\kappa}_1 = -9.060 \ (\bar{v}_1 = 0.92)$$
$$\bar{\kappa}_2 = -25.76 \ (\bar{v}_2 = 2.61)$$

The first one is practically the same of the preceding case, showing how, for the translation failure mode that is related to the sum of the translation stiffness of the two columns, the greater flexibility of the more compressed column is compensated by the lower flexibility of the less compressed column. The second eigenvalue is much lower due to the higher compression of the right column that fails independently from the other.

Omitting the calculation details, that are the same shown in the preceding case, the two eigenvectors related to κ_1 and κ_2 are eventually reported:

$$\begin{vmatrix} +0.203 \\ +0.190 \\ +1.000 \end{vmatrix} \quad \begin{vmatrix} +0.879 \\ -1.000 \\ -0.008 \end{vmatrix}$$

One notices a prevalent translation failure mode for the first eigenvalue and a prevalent rotation failure mode for the second eigenvalue.

In comparison with the general method that requires a repeated evaluation of the determinant, as shown in the simple example presented above, the linearized theory that has the unknown parameter γ (or ν) in binomial forms such as

$$k = k_o + \gamma \Delta k$$

allows the adoption of more efficient solution procedures, particularly convenient when the eigenvalues of large dimensions matrices are to be calculated. Many computer codes use systematically these procedures of linear algebra.

Without entering the complex topics of the numerical calculation, the approximations of the second-order linearized theory are to be reminded, so that a correct use of such procedures is made. First one has to keep in mind that the instability condition is applied to a set of equilibrium equations of which, following the Displacement Method, the unknowns are the displacement components of the joints. With this premise, one has that:

- the situations less restrained that reach the instability limit with high values of the joint displacements lead to much more accurate results than the more restrained situations with low values of the joint displacements;
- at the limit of the double fixed end support with null joint displacements, for which the correct theory leads to undetermined expression such as $\infty - \infty$, the linearized theory instead leads to a determined solution that doesn't perceive the instability of the member;
- at a parity of the other conditions, the linearized theory leads to good results for behaviours with prevalent translations, while it is less reliable for behaviours with prevalent rotations,
- beyond the limit $\nu = -1$, the behaviours with prevalent rotations cannot be treated with the linear approximation;
- a simple contrivance to move the configuration of the set of equilibrium equations into the field of less restrained situations with prevalent translation behaviour consists of introducing additional calculation joints within the length of the members.

Fig. 4.33 Column with
different restraint degrees

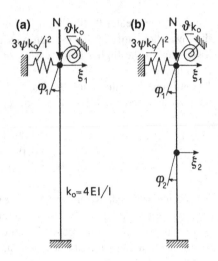

$$k_o = 4EI/l$$

In particular this last contrivance, although enlarges the solving set of equations
with consequent increase of the calculation burden, allows to apply, with good reliability of results, the linearized theory to all types of structure.

In order to give a more complete indication on the result differences in comparison
with the right ones of the correct theory, the instability condition has been elaborated
for the simple column of Fig. 4.33a. There is one only calculation joint, on which
the two unknowns φ_1, ξ_1 have been evidenced and for which the two equilibrium
equations $m_1 = 0$, $r_1 = 0$ have been written. The elastic constants ϑ, ψ of the
springs are expressed as a ratio to that $k_o = 4EI/l$ of the column. High values of these
constants lead to very restrained situations and vice versa. By varying their ratio ϑ/ψ,
the translation behaviour can be privileged in comparison with the rotation one or
vice versa.

The following table gives the values of the critical load calculated with the linearized theory as a ratio to those of the correct theory. One can notice how the
approximations are always in excess that is how the linearized coefficients interpret
the members as more rigid. One can notice again how the errors grow greatly with
the higher support stiffness and with the lower joint translations.

ϑ	0.0	0.5	1.0	2.0	4.0	8.0
ψ						
0.0	1.004	1.005	1.005	1.005	1.005	1.006
0.5	1.015	1.012	1.009	1.008	1.008	1.009
1.0	1.034	1.022	1.016	1.013	1.013	1.013
2.0	1.104	1.086	1.050	1.025	1.023	1.023
4.0	1.196	1.314	1.393	1.248	1.192	1.052
8.0	1.214	1.338	1.461	1.619	1.593	1.510

The same column has been analysed also with an additional calculation joint (see Fig. 4.33b) that is with the two additional unknowns φ_2, ξ_2 and with four equilibrium equations. Again with reference to the parameters ϑ, ψ of rotation and translation stiffness of the springs, in the following table are reported the new results, that show a very good and uniform precision of the linearized theory over all the range examined.

ϑ	0.0	0.5	1.0	2.0	4.0	8.0
ψ						
0.0	1.000	1.001	1.002	1.003	1.003	1.003
0.5	1.010	1.008	1.007	1.006	1.006	1.006
1.0	1.012	1.010	1.008	1.006	1.006	1.006
2.0	1.012	1.010	1.008	1.006	1.006	1.006
4.0	1.013	1.010	1.008	1.006	1.006	1.006
8.0	1.013	1.010	1.008	1.006	1.006	1.006

4.4 Examples of Second-Order Analysis

A discourse is stated beforehand about the behaviour of the frame structures in which the bracing functions versus horizontal actions are concentrated on a special post of greater stiffness. Let's consider for example the simplified case of the frame of Fig. 4.34a, where there is the current column "b", with null flexural stiffness, charged with the resistance versus the vertical loads P_b by simple axial action. Its instability verification will be made with reference to the elementary scheme of Euler's column, with the buckling length equal to its height ($l_o = h$). The top of the current column is connected, through the beam "a", to the post "c" that, with its high translation flexural stiffness, ensures the overall equilibrium of the structure.

So the post "c" is the bracing element versus the horizontal actions that, in the concerned case (see Fig. 4.34), are:

- the explicit force H due for example to the wind pressure;
- the deviatory action due to the unintended eccentricity and related to the possible verticality error of the column "b";
- the second-order deviatory action due to the not negligible value of the translation ξ.

To these horizontal actions, the one deriving from possible shear forces in the columns may be added, due to flexural components of the internal axial action or, in the case of columns connected with full (moment resisting) supports to the beams, to the hyperstatic moments exchanged with the beams.

For what concerns the verticality error e, it comes from the execution tolerance and can be quantified on the basis of the type of construction as a ratio to the height of the column:

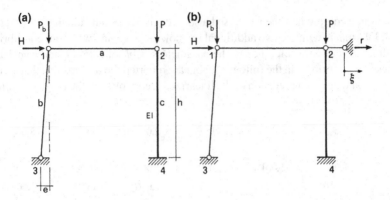

Fig. 4.34 Analysis of the type structure

$$\varepsilon = \frac{e}{h}$$

for example within $\varepsilon = 1/1000$ and $\varepsilon = 1/500$.

The structure is then analysed assuming, with the Displacement Method procedure, the translation ξ as the unknown of the problem (see Fig. 4.34b) and setting the equilibrium equation of the beam ($r = 0$).

The second-order translation corrected stiffness of the member with a full and a hinged end support (see Sect. 4.2.1) are adopted:

$$k'_{vi} = \frac{3}{EI}h^2 G'$$

$$k'_v = \frac{3}{EI}h^3 G'_v$$

where with $\alpha^2 = P/EI$, it is:

$$G' = G'(\alpha h) = \frac{1}{g_1(\alpha h)}$$

$$G'_v = G'_v(\alpha h) = \left\{ \frac{1}{g1(\alpha h)} - \frac{(\alpha h)^2}{3} \right\}$$

and the instability function is:

$$g_1 = g_1(\alpha h) = \frac{3}{\alpha h}\left\{ \frac{1}{\alpha h} - \frac{1}{tg\alpha h} \right\}$$

For the equilibrium, the reaction of the additional translation support is null:

$$\frac{3EI}{h^3}G'_v\xi - H - P_b\frac{e}{h} - P_b\frac{\xi}{h} = 0$$

In this equation, the first term represents the elastic reaction of the post "c" subjected to the top translation ξ, the second term represents the explicit horizontal action, the third term represents the deviatory effect of the vertical load on the out of plumb column, the forth term represents the second-order effect of the vertical load on the same column "b".

The solution is:

$$\xi = \frac{H + P_b e/h}{\dfrac{3EI}{h^3}G'_v - \dfrac{P_b}{h}}$$

and by consequence the base moment of the post is:

$$M_4 = \frac{3EI}{h^2}G'\xi = \frac{G'}{G'_v - \dfrac{P_b h^2}{3EI}}(Hh + P_b e) = \gamma M_o$$

With $M_o = Hh + P_b e$ the first-order contribution is indicated which is proportional to the applied actions; with γ the amplification factor is indicated due to the second-order effects that increase the base moment all the more so higher are the vertical loads. Introducing the expressions of G' and G'_v one has:

$$\gamma = \frac{\dfrac{1}{g_1}}{\dfrac{1}{g_1} - \dfrac{(\alpha h)^2}{3} - \dfrac{P_b h^2}{3EI}}$$

In the case of several current columns (see Fig. 4.35), in P_b the sum of all the related vertical loads is to be put:

$$P_b = \sum P_i$$

The expression of γ contains the contribution of the current pendulum columns to the second-order effects on the bracing post. Also if this latter has no any axial load ($P = 0$ with $\alpha h = 0$ and $g_1 = 1$), the second-order effects coming from the other columns remain with:

$$\gamma = \frac{1}{1 - \dfrac{P_b h^2}{3EI}}$$

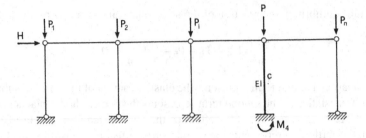

Fig. 4.35 Multi-bay structure

and the possible instability can be reached when $\gamma \to \infty$, that is when:

$$\frac{P_b h^2}{3EI} = 1$$

with a critical load

$$P_{bE} = \frac{3EI}{h^2}$$

Introducing in the original equation (with $P > 0$) the expression of g_1, one has:

$$\gamma = \frac{1}{1 - \dfrac{3}{\alpha h}\left(\dfrac{1}{\alpha h} - \dfrac{1}{tg\alpha h}\right)\dfrac{(P + P_b)h^2}{3EI}} = \frac{1}{1 - \left\{1 - \dfrac{\alpha h}{tg\alpha h}\right\}(1 + c)}$$

with

$$c = \frac{P_b}{P}$$

Figure 4.36 shows the curves of this second-order amplification factor $\gamma = \gamma(\alpha h, c)$. The instability limit, corresponding to $\gamma = \infty$, is defined by the condition:

$$\left(1 - \frac{\alpha h}{tg\alpha h}\right)(1 + c) = 1$$

that leads to:

$$tg\alpha h = \frac{1 + c}{c}\alpha h$$

With respect to the solution of the isolated column (with $c = 0$), that is deduced from

Fig. 4.36 Curves of the
amplification factor

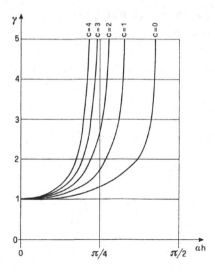

$$\text{tg}\,\alpha\,h = +\infty$$

and gives:

$$\overline{\alpha}h = \frac{\pi}{2} \quad (\beta = \frac{\pi}{\overline{\alpha}h} = 2)$$

the presence of current pendulum columns connected to the bracing post reduces the
critical load P'_E as indicated in Fig. 4.37 where the solution of the instability condition
is graphically represented.

For $c = P_b/P = 1$ with $(1 + c)/c = 2$, one obtains for example:

$$\overline{\alpha}h = 0.37\pi$$

with

$$\beta = \frac{\pi}{\overline{\alpha}h} = 2.70$$

that gives a buckling length:

$$l_0 = 2.70\,h$$

much greater that the one of the isolated column and a critical load much lower:

$$P'_E = \frac{P_E}{\beta^2} \simeq \frac{P_E}{7.3} \quad (\text{instead of } P_E/4)$$

Fig. 4.37 Critical load P'_E

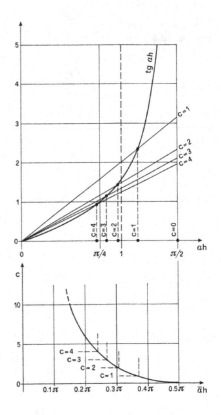

referred to the axial load P on the post itself.

What stated above shows how, for the stability of the bracing elements of the buildings, on which the resistance versus the horizontal actions is concentrated, much larger dimensions are needed than what derives from their height. The second-order effects received from the connected current columns increase very much the deviatory actions on the bracing element, requiring for this latter a big stiffness and consequent slenderness ratios much lower than those of the ordinary current or isolated columns.

4.4.1 Multi-storey Braced Continuous Column

The continuous column of Fig. 4.38, belonging to a common typology of braced frames, is to be analysed in order to evaluate the instability parameters for the pertinent verifications. The column segments are assumed of equal height and equal cross section "EI". From any floor the same vertical load is transmitted with:

$$P_a = P_b = P_c = P$$

Fig. 4.38 Continuous
column

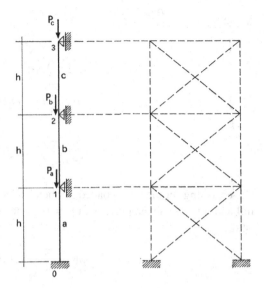

and consequently the axial actions are:

$$N_c = 1P = \frac{1}{3}N$$

$$N_b = 2P = \frac{2}{3}N$$

$$N_a = 3P = N$$

With reference to the internal joints 1 and 2, the equilibrium equations of the Displacement Method are set with:

$$\begin{cases} \left(\dfrac{4EI}{h}G_{1a} + \dfrac{4EI}{h}G_{1b}\right)\varphi_1 + \dfrac{2EI}{h}G_{ib}\varphi_2 = 0 \\[3mm] \dfrac{2EI}{h}G_{ib}\varphi_1 + \left(\dfrac{4EI}{h}G_{1b} + \dfrac{3EI}{h}G_c'\right)\varphi_2 = 0 \end{cases}$$

where in particular it is $G_c' = 1/g_{1c}$.

The instability condition is set equating to zero the determinant of the coefficient matrix:

$$4(G_{1a} + G_{1b})\left(G_{1b} + \frac{3}{4}G_c'\right) - G_{ib}^2 = 0$$

Assuming now:

$$\alpha_a = \sqrt{\frac{N}{EI}} = \alpha$$

$$\alpha_b = \sqrt{\frac{2}{3}}\alpha = 0.8165\alpha$$

$$\alpha_c = \sqrt{\frac{1}{3}}\alpha = 0.5774\alpha$$

and expecting for the more loaded segment "a" a behaviour near to the condition of full and hinged end supports with $\beta = 0.7$, the solution is searched for around the value:

$$\alpha h = \frac{\pi}{\beta} = 1.43\pi$$

– First trial with $\alpha h = 1.42\pi$:

$$\alpha_a h = 1.4200\pi \qquad \alpha_b h = 1.1594\pi \qquad \alpha_c h = 0.8199\pi$$
$$G_{1a} = 0.0230 \qquad G_{1b} = 0.4492 \qquad G_{ib} = 1.3643$$
$$G'_c = 0.4375 \qquad det = -0.3929$$

– Second trial with $\alpha h = 1.38\pi$:

$$\alpha_a h = 1.3800\pi \qquad \alpha_b h = 1.1268\pi \qquad \alpha_c h = 0.7968\pi$$
$$G_{1a} = 0.1067 \qquad G_{1b} = 0.4878 \qquad G_{ib} = 1.3327$$
$$G'_c = 0.4777 \qquad det = +0.2356$$

– Third trial with:

$$\alpha h = 10.38\pi + 0.04\pi\frac{0.2356}{0.2356 + 0.3929} = 1.395\pi$$

$$\alpha_a h = 1.3950\pi \qquad \alpha_b h = 1.1390\pi \qquad \alpha_c h = 0.8054\pi$$
$$G_{1a} = 0.0764 \qquad G_{1b} = 0.4736 \qquad G_{ib} = 1.3442$$
$$G'_c = 0.4629 \qquad det = -0.0013 (\approx 0)$$

So a support factor:

$$\beta_a = \frac{\pi}{1.3950\pi} = 0.72$$

is obtained, very near to the one previously expected for the segment "a", with negligible continuity effects exchanged with the upper segment "b", being this latter less loaded but also less restrained.

For the other segments, one has:

$$\beta_b = \frac{\pi}{1.1390\pi} = 0.88$$

$$\beta_c = \frac{\pi}{0.8054\pi} = 1.24$$

The effects of the continuity can be noticed that gives greater restraints to the more loaded segment "b" at the expense of the upper less loaded segment "c".

4.4.2 A Multi-portal Frame

Let's consider the multi-portal frame of Fig. 4.39 made, following a widespread scheme of the precast constructions, of a line of n columns connected at their top with hinged beams. On the internal columns a load P is applied, on the end columns a halved load is applied.

With reference to the beam line, the equation of translation equilibrium is set as:

$$\left\{ 2\frac{3EI}{h^3}G'_{v0} + (n-2)\frac{3EI}{h^3}G'_{v1} \right\}\xi = H$$

having assumed all the columns with the same cross section "EI" and having indicated with the subscripts 0 and 1, respectively, the quantities related to the end columns (with a load $P/2$) and the internal columns (with a load P).

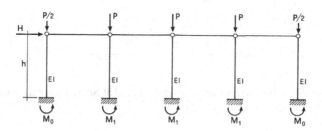

Fig. 4.39 Multi-portal frame

The instability condition is set with:

$$2G'_{v0} + (n - 2)G'_{v1} = 0$$

Of this equation, with:

$$\alpha_1 = \sqrt{\frac{P}{EI}} = \alpha \qquad \alpha_0 = \sqrt{\frac{1}{2}}\alpha = 0.7071\alpha$$

and $\beta < 2$, the root:

$$\alpha h = \frac{\pi}{\beta} > \pi/2$$

is searched for, expecting, due to the frame connection, a greater restraint given to the more loaded internal columns by the two end less loaded columns.

For numerical calculation facility, in the range of possible use, a table is given here below related to the stiffness correction functions of the member with a full and a hinged end supports:

$$k'_1 = \frac{3EI}{h}G'(\alpha h)$$

$$k'_{vi} = k'_1/h$$

$$k'_v = \frac{3EI}{h^3}G'_v(\alpha h)$$

where the functions:

$$G'(\alpha h) = \frac{1}{g_1} = \frac{4G_1^2 - G_i^2}{3G_1}$$

$$G'_v(\alpha h) = \frac{1}{g_1} - \frac{(\alpha h)^2}{3} = \frac{4G_1G_v - 3G_{vi}^2}{G_1}$$

can be evaluated on the basis of those already tabled at Sect. 4.3 for the member with two full end supports.

$\alpha h/\pi$	G'	G'_v
0.00	1.0000	1.0000
0.02	0.9997	0.9984
0,04	0.9989	0.9937
0.06	0.9976	0.9858
0.08	0.9958	0.9747
0.10	0.9934	0.9605
0.12	0.9905	0.9431
0.14	0.9870	0.9226
0.16	0.9830	0.8988
0.18	0.9785	0.8719
0.20	0.9734	0.8418
0.22	0.9677	0.8085
0.24	0.9615	0.7720
0.26	0.9546	0.7323
0.28	0.9472	0.6893
0.30	0.9392	0.6431
0.32	0.9306	0.5937
0.34	0.9213	0.5410
0.36	0.9114	0.4851
0.38	0.9009	0.4258
0.40	0.8896	0.3632
0.42	0.8777	0.2973
0.44	0.8650	0.2281
0.46	0.8516	0.1555
0.48	0.8374	0.0794
0.50	0.8225	0.0000
0.52	0.8067	−0.0829
0.54	0.7900	−0.1693
0.56	0.7725	−0.2592
0.58	0.7540	−0.3527
0.60	0.7345	−0.4499
0.62	0.7140	−0.5506
0.64	0.6924	−0.6551
0.66	0.6697	−0.7634
0.68	0.6458	−0.8755
0.70	0.6206	−0.9915
0.72	0.5940	−1.1115
0.74	0.5660	−1.2356
0.76	0.5364	−1.3638
0.78	0.5052	−1.4964
0.80	0.4722	−1.6333
0.82	0.4372	−1.7749
0.84	0.4002	−1.9211
0.86	0.3609	−2.0723
0.88	0.3192	−2.2285
0.90	0.2747	−2.3901

(continued)

(continued)

$\alpha h/\pi$	G'	G'_v
0.92	0.2272	−2.5574
0.94	0.1764	−2.7305
0.96	0.1219	−2.9100
0.98	0.0633	−3.0963
1.00	0.0000	−3.2899

Figure 4.40 points out the intersections of the curve:

$$G'_{v0} = G'_v(\alpha h/\sqrt{2})$$

with the curves:

$$\frac{2-n}{2}G'_{v1} = \frac{2-n}{2}G'_v(\alpha h)$$

when the parameter n varies. These intersections give the solutions of the instability condition when the number of columns varies.

In Fig. 4.41, these results are reported in terms of the support factors:

$$\beta_0 = \beta_0(n) = \sqrt{2}\pi/\overline{\alpha}h$$

Fig. 4.40 Graphic representation of the solution

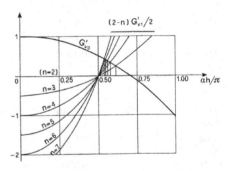

Fig. 4.41 Support factors values

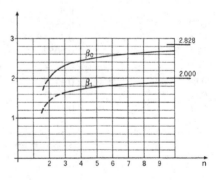

$$\beta_1 = \beta_1(n) = \pi/\overline{\alpha}h$$

respectively of the end columns and the internal ones. The formers are to be verified under the load $P/2$, the latters under the load P.

To be noticed how, as the number of columns increases, the effectiveness of the end less loaded ones decreases and the situation approaches to that of all equal current columns behaving like isolated members ($\beta_1 = 2$). With three columns only the inner one, with a support factor $\beta_1 = 1.61$, receives the greater stabilizing contribution from the outer ones. The end columns start from a support factor $\beta_0 = 2$ for $n = 2$ when, in the absence of inner more loaded columns, their behaviour is equal to that of the isolated column, and then go up to ever greater values approaching asymptotically to the limit $\beta_0 = 2/\sqrt{2} = 2.83$.

Also for the stiffness correction functions of the member with a full and a hinged end supports, it is possible to provide the approximate linearized expressions similar to those presented at Sect. 4.3.2 for the member with two full end supports. So one has:

$$F'(\kappa) = G'(\kappa) \simeq 1 + \frac{1}{15}\kappa \simeq 1 + 2v/3$$

$$F'_v(\kappa) = G'_v(\kappa) \simeq 1 + \frac{2}{5}\kappa \simeq 1 + 4v$$

with $\kappa = (\alpha h)^2$ and $v = N/N_E \approx \kappa/10$. For these expressions, the same comments stated at Sect. 4.3.2 are valid, with a much higher precision of the translation linearized function with respect to the other one. In particular, the linearization of the rotation function G' leads to very big errors for $v < -0.5$.

With the linearized expressions, the equilibrium equation of the multi-portal of concern (of Fig. 4.39) becomes:

$$\left\{ 2\frac{3EI}{h^3}(1 + 4v_0) + (n - 2)\frac{3EI}{h^3}(1 + 4v_1) \right\}\xi = H$$

and leads to the solution (with $v_0 = cv$, $v_1 = v$ and $c = N_0/N_1$):

$$\xi = \left(\frac{h^3}{3EI} \frac{H}{n} \right) \frac{n}{2(1 + 4cv) + (n - 2)(1 + 4v)}$$

having evidenced in brackets the first-order value. The correction denominator related to the second-order contribution can be reduced to:

$$d(v; n) = [n + 4v(2c + n - 2)]/n$$

From this expression, one can immediately deduce, from the instability condition $d = 0$, the explicit form of the critical load:

$$-\bar{v} = \frac{n}{4(2c + n - 2)} = \bar{v}_1 \quad (\bar{v}_0 = c\bar{v})$$

obtaining for example with very good precision the same curves:

$$\beta_1 = 1/\sqrt{-\bar{v}_1} \quad \beta_0 = 1/\sqrt{-\bar{v}_0}$$

drawn for $c = 1/2$ in Fig. 4.41.

One can also evaluate in a very simple way the bending moment at the base of the columns with:

$$M_0 = \frac{3EI}{h^2} G_0'\xi = \left[\frac{Hh}{n}\right] \frac{n(1 + 2cv/3)}{n + 4v(2c + n - 2)}$$

$$M_1 = \frac{3EI}{h^2} G_1'\xi = \left[\frac{Hh}{n}\right] \frac{n(1 + 2v/3)}{n + 4v(2c + n - 2)}$$

With $c = 1/2$ and assuming a level of the axial action equal to:

$$v = -0.150(= v_1) \quad (v_0 = 0.075)$$

one has for example the base moments calculated here below.

– For $n = 3$:

$$\xi = \left[\frac{h^3}{3EI} \frac{H}{n}\right] \frac{n}{n + 4v(n - 1)} = \left[\frac{h^3}{3EI} \frac{H}{n}\right] 1.667$$

$$M_0 = \left[\frac{Hh}{n}\right] 1.667(1 + v/3) = \left[\frac{Hh}{n}\right] 1.583$$

$$M_1 = \left[\frac{Hh}{n}\right] 1.667(1 + 2v/3) = \left[\frac{Hh}{n}\right] 1.500$$

– For $n \to \infty$:

$$\xi = \left[\frac{h^3}{3EI} \frac{H}{n}\right] \frac{1}{1 + 4v} = \left[\frac{h^3}{3EI} \frac{H}{n}\right] 2.500$$

$$M_0 = \left[\frac{Hh}{n}\right] 2.500(1 + v/3) = \left[\frac{Hh}{n}\right] 2.375$$

$$M_1 = \left[\frac{Hh}{n}\right] 2.500(1 + 2v/3) = \left[\frac{Hh}{n}\right] 2.250$$

One can notice the large amplifications of the moments due to the vertical loads acting on the deformed drifted configuration of the structure, with geometric and static effects greater than 1.5 times the first-order ones for the more stable case of three columns, larger than 2 times the first-order ones for the less stable case of many columns. In this latter case, the behaviour of the internal columns is the current one, without reciprocal help, corresponding to what evaluated for the isolated column at Sect. 4.1.3. With $v = -0.15$, the "exact" equations obtained in that section give (with $\alpha h = \pi\sqrt{-v} = 1.2167$):

$$\xi = \left[\frac{h^3}{3EI}\frac{H}{n}\right]\left\{\frac{3}{(\alpha h)^3}(\text{tg}\alpha h - \alpha h)\right\} = \left[\frac{h^3}{3EI}\frac{H}{n}\right] 2.479$$

$$M_1 = \left[\frac{Hh}{n}\right]\frac{\text{tg}\alpha h}{\alpha h} = \left[\frac{Hh}{n}\right] 2.223$$

Should be taken into account that the level of the axial action here assumed, compared to the critical load $v'_E = -0.25$ of the cantilever column, corresponds to the:

$$\frac{v}{v'_E} = \frac{0.15}{0.25} = 0.6$$

of the ultimate bearing capacity. This is a high level, such as those that are analysed under the design values of the actions amplified by γ_F for the purposes of the verifications of the resistance limit states.

4.4.3 Multi-storey Frame

Let's consider the two storeys frame off Fig. 4.42, made of the two beams "a" and "b" with constant cross section of stiffness EI' and of the columns "c"–"d" at first floor and "e"–"f" at ground floor all with constant cross section or stiffness EI. For the evaluation of the second-order effects, the following axial actions are assumed:

$$N_a = N_b = 0$$
$$N_c = N_d = -pl/2 = N/2$$
$$N_e = N_f = -2pl/2 = N$$

Reference can be made to Sects. 4.3.1 and 4.3.3 for what concerns the influence of this approximate evaluation on the results of the structural analysis.

Fig. 4.42 Two storeys frame

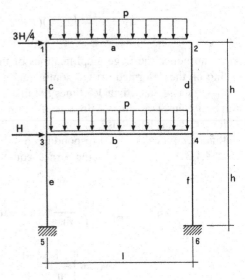

Fig. 4.43 Halved structures for symmetric and anti-symmetric analyses

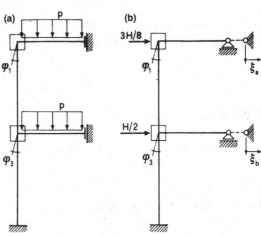

First the symmetric behaviour of the frame without horizontal actions ($H = 0$) is analysed. For this case, with reference to the halved structure of Fig. 4.43a, the geometric unknowns φ_1 and φ_3 are assumed. By consequence, the two rotation equilibrium equations are set:

$$
\begin{cases}
\left[\dfrac{2EI'}{l} + \dfrac{4EI}{h}G_{1c} \right]\varphi_1 + \dfrac{2EI}{h}G_{ic}\varphi_3 = \dfrac{pl^2}{12} \\[4mm]
\dfrac{2EI}{l}G_{ic}\varphi_1 + \left[\dfrac{2EI'}{l} + \dfrac{4EI}{h}G_{1c} + \dfrac{4EI}{h}G_{1e} \right]\varphi_3 = \dfrac{pl^2}{12}
\end{cases}
$$

Of these equations, the solution is searched for the following data:

$$1 = 8h/5$$
$$I' = 4I$$
$$v = N/N_E = -0.5 \quad (v_e = 2v_c)$$

So, with $\overline{M} = pl^2/12$, the equations become.

$$\begin{cases} [5 + 4G_1(v/2)]\varphi_1 + [2G_i(v/2)]\varphi_3 = \dfrac{\overline{M}h}{EI} \\[3mm] [2G_i(v/2)]\varphi_1 + [5 + 4G_1(v/2) + G_1(v)]\varphi_3 = \dfrac{\overline{M}h}{EI} \end{cases}$$

where the correction functions are here evaluated with the approximate linear expressions (see Sect. 4.3.2):

$$G_{1c} = 1 + v/6 = 0.9167$$
$$G_{ic} = 1 - v/12 = 1.0417$$
$$G_{1e} = 1 + v/3 = 0.8333$$
$$G_{ie} = 1 - v/6 = 1.0833$$

The equations eventually become:

$$\begin{cases} 8.6668\varphi_1 + 2.0834\varphi_3 = \dfrac{\overline{M}h}{EI} \\[4mm] 2.0834\varphi_1 + 12.000\varphi_3 = \dfrac{\overline{M}h}{EI} \end{cases}$$

and lead to the solution:

$$\varphi_1 = 0.0995\frac{\overline{M}h}{EI} \quad (= -\varphi_2)$$

$$\varphi_3 = 0.0661\frac{\overline{M}h}{EI} \quad (= -\varphi_4)$$

$$(\xi_a = \xi_b = 0)$$

from which the moments are calculated as:

$$M_{a1} = -\overline{M} + \frac{2EI'}{1}\varphi_1 = -0.5025\overline{M}$$

$$M_{a0} = +M_{a1} + \frac{3}{2}\overline{M} = +0.9975\overline{M}$$

$$M_{c1} = -\frac{4EI}{h}G_{1c}\varphi_1 - \frac{2EI}{h}G_{ic}\varphi_3 = -0.5025\overline{M}$$

$$M_{c3} = +\frac{2EI}{h}G_{ic}\varphi_1 + \frac{4EI}{h}G_{1c}\varphi_3 = +0.4495\overline{M}$$

$$M_{b3} = -\overline{M} + \frac{2EI'}{1}\varphi_3 = -0.6695\overline{M}$$

$$M_{b0} = +M_{b3} + \frac{3}{2}\overline{M} = +0.8305\overline{M}$$

$$M_{e3} = -\frac{4EI}{h}G_{1e}\varphi_3 = -0.2200\overline{M}$$

$$M_{e5} = +\frac{2EI}{h}G_{ie}\varphi_3 = +0.1432\overline{M}$$

having assumed positive the moments that tension the lower side of the beams and the right side of the columns.

Figure 4.44 shows for the entire structure the diagram of the bending moment drawn with a continuous line. It can be noticed that the first-order solution:

$$M_{a1} = -0.513\overline{M} \qquad M_{a0} = +0.987\overline{M}$$
$$M_{c1} = -0.513\overline{M} \qquad M_{c3} = +0.442\overline{M}$$
$$M_{b3} = -0.690\overline{M} \qquad M_{b0} = +0.810\overline{M}$$
$$M_{e3} = -0.278\overline{M} \qquad M_{e5} = +0.139\overline{M}$$

Fig. 4.44 Diagram of bending moment

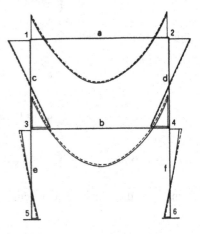

drawn with a dashed line in the same figure, obtained with unity values of the correction functions, is, for this symmetric rotation behaviour, very near.

The second case to be examined, again with reference to the frame of Fig. 4.42, is the anti-symmetric behaviour without vertical loads. This analysis is performed in order to examine the effects of the wind pressure separately from those of the contemporary vertical loads. The effects of the two types of loads shall be then summed up, but this can be done under the same axial actions as explained in the introduction of Sect. 4.1. The same data of the preceding case are referred to also for the halved structure of Fig. 4.43b, where the simplifications due to the anti-symmetry have been introduced.

On the basis of the four geometric unknowns φ_1, φ_3, ξ_a, ξ_b of the problem evidenced in the figure, the solving algorithm is set up, made of two rotation equilibrium equations and two translation equilibrium equations:

$$
\begin{cases}
m_{11}\varphi_1 + m_{13}\varphi_3 + m_{1a}\xi_a + m_{1b}\xi_b = 0 \\
m_{31}\varphi_1 + m_{33}\varphi_3 + m_{3a}\xi_a + m_{3b}\xi_b = 0 \\
r_{a1}\varphi_1 + r_{a3}\varphi_3 + r_{aa}\xi_a + r_{ab}\xi_b = 3H/8 \\
r_{b1}\varphi_1 + r_{b3}\varphi_3 + r_{ba}\xi_a + r_{bb}\xi_b = 1H/2
\end{cases}
$$

where

$$
m_{11} = \frac{6EI'}{1} + \frac{4EI}{h}G_{1c}
$$

$$
m_{13} = m_{31} = \frac{2EI}{h}G_{ic}
$$

$$
m_{33} = \frac{6EI'}{1} + \frac{4EI}{h}G_{1c} + \frac{4EI}{h}G_{1e}
$$

$$
m_{1a} = r_{a1} = -\frac{6EI}{h^2}G_{vic} = m_{3a} = r_{a3}
$$

$$
m_{1b} = r_{b1} = +\frac{6EI}{h^2}G_{vic}
$$

$$
m_{3b} = r_{b3} = +\frac{6EI}{h^2}G_{vic} - \frac{6EI}{h^2}G_{vc}
$$

$$
r_{aa} = \frac{12EI}{h^3}G_{vc} = -r_{ab} = -r_{ba}
$$

$$
r_{bb} = \frac{12EI}{h^3}G_{vc} + \frac{12EI}{h^3}G_{ve}
$$

The translation correction functions of the columns are evaluated, like the rotation ones already used, with the approximate linearized expressions (with $\nu = -0.5$):

$$
G_{vic} = 1 + \nu/12 = 0.9875
$$

$$G_{vc} = 1 + v/2 = 0.7500$$
$$G_{vie} = 1 + v/6 = 0.9167$$
$$G_{ve} = 1 + v = 0.5000$$

With these values, the equations become:

$$\begin{cases} +18.6668\varphi_1 + 2.0834\varphi_3 - 5.9250\bar{\xi}_a + 5.9250\bar{\xi}_b = 0 \\ +2.0834\varphi_1 + 22.0000\varphi_3 - 5.9250\bar{\xi}_a + 0.4248\bar{\xi}_b = 0 \\ -5.9250\varphi_1 - 5.9250\varphi_3 + 9.0000\bar{\xi}_a - 9.0000\bar{\xi}_b = 3Hh^2/(8EI) \\ +5.9250\varphi_1 + 0.4248\varphi_3 - 9.0000\bar{\xi}_a + 15.0000\bar{\xi}_b = 1Hh^2/(2EI) \end{cases}$$

where $\bar{\xi}_a = \xi_a/h$ and $\bar{\xi}_b = \xi_b/h$ have been set. With $\overline{M}' = Hh/2$ the solution is:

$$\varphi_1 = 0.0541\frac{\overline{M}'h}{EI} \quad (=\varphi_2)$$

$$\varphi_3 = 0.1682\frac{\overline{M}'h}{EI} \quad (=\varphi_4)$$

$$\bar{\xi}_a = 0.6756\frac{\overline{M}'h}{EI}$$

$$\bar{\xi}_b = 0.4459\frac{\overline{M}'h}{EI}$$

from which the following moments are obtained:

$$M_{a1} = +\frac{6EI'}{l}\varphi_1 = +\frac{15EI}{h}\varphi_1 = +0.8120\overline{M}'$$

$$M_{c1} = -\frac{4EI}{h}G_{1c}\varphi_1 - \frac{2EI}{h}G_{ic}\varphi_3 + \frac{6EI}{h}G_{vic}(\bar{\xi}_a - \bar{\xi}_b) = +0.8120\overline{M}'$$

$$M_{c3} = +\frac{2EI}{h}G_{ic}\varphi_1 + \frac{4EI}{h}G_{1c}\varphi_3 - \frac{6EI}{h}G_{vic}(\bar{\xi}_a - \bar{\xi}_b) = -0.6314\overline{M}'$$

$$M_{b3} = +\frac{6EI'}{l}\varphi_3 = +\frac{15EI}{h}\varphi_3 = +2.5232\overline{M}'$$

$$M_{e3} = -\frac{4EI}{h}G_{1e}\varphi_3 + \frac{6EI}{h}G_{vie}\bar{\xi}_b = +1.8918\overline{M}'$$

$$M_{e5} = +\frac{2EI}{h}G_{ie}\varphi_3 - \frac{6EI}{h}G_{vie}\bar{\xi}_b = -2.0880\overline{M}'$$

expressed with the same sign conventions quoted before. Figure 4.45 shows for the entire structure the diagram of the bending moment drawn with a continuous line.

The combination of this diagram with that of Fig. 4.44 obviously depends on the ratio p/H between the intensities of the related loads.

From this solution, that gives not very different moments at the base and the top of the columns, one can notice the big stiffness of the beams compared to that of the columns, with prevalence of the translation (floor drifts) effects with respect to the joint rotation effects. For beams with infinite flexural stiffness, the first-order behaviour would give (see Fig. 4.43b):

$$M_{c1} = -M_{c3} = \frac{3H}{8}\frac{h}{2} = +0.3750\overline{M}'$$

$$M_{e3} = -M_{e5} = \frac{7H}{8}\frac{h}{2} = +0.8750\overline{M}'$$

The related bending moment diagram is indicated with the dotted lines in Fig. 4.45. The dashed lines correspond instead to the correct first-order solution:

$$\begin{array}{ll} M_{a1} = +0.422\overline{M}' & M_{b3} = +10.128\overline{M}' \\ M_{c1} = +0.422\overline{M}' & M_{c3} = -0.328\overline{M}' \\ M_{e3} = +0.800\overline{M}' & M_{e5} = -0.950\overline{M}' \end{array}$$

obtained from the equations written before with unity values for the correction functions.

It can be seen how the second-order deviation effects of the vertical loads, acting on the deformed drifted configuration of the frame, lead to more than doubled moments. This is due to the high axial actions with $v = -0.5$ that can correspond to the design values, amplified by γ_F for the verifications of the resistance limit states.

The results of the analysis would be somewhat different if applied with more accurate evaluations of the axial actions, taking into account, for example, the dis-

Fig. 4.45 Diagrams of bending moments

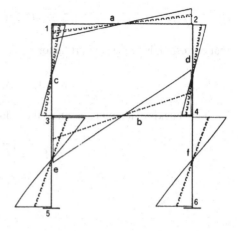

symmetry of these actions between the left and the right columns, being the formers less compressed and the latters more compressed due to the wind pressure. But these differences would have a very small influence on the critical load, as shown at Sect. 4.3.3 (see Fig. 4.32).

So at the end of this example, the calculation of the critical load is presented, using again the linearized expressions of the stiffness in the equations written for the analysis under the horizontal actions. The terms of the coefficient matrix, having simplified EI/h, are expressed by:

$$c_{11} = 15 + 4(1 + v/6)$$
$$c_{12} = c_{21} = 2(1 - v/12)$$
$$c_{13} = c_{31} = -6(1 + v/12)$$
$$c_{14} = c_{41} = +6(1 + v/12)$$
$$c_{22} = 15 + 4(1 + v/6) + 4(1 + v/3)$$
$$c_{23} = c_{32} = -6(1 + v/12)$$
$$c_{24} = c_{42} = +6(1 + v/12) - 6(1 + v/6)$$
$$c_{33} = 12(1 + v/2)$$
$$c_{34} = c_{43} = -12(1 + v/2)$$
$$c_{44} = 12(1 + v/2) + 12(1 + v)$$

The instability condition:

$$\det[C(v)] = 0$$

is searched for around $v = -1$, value that corresponds, for the column "e" fully loaded, to the support condition of two end rotation restraints ($\beta = 1$).

Figure 4.46 shows the determinant curve evaluated in the interval $-0.80 \geq v \geq -1.00$. The intersection provides the solution:

$$\bar{v} = -0.865 \quad (= v'_E)$$

that corresponds to the support factor:

$$\beta = 1/\sqrt{-\bar{v}} = 1.08$$

With the buckling length $h_o = \beta h$ of the column "e", the critical load becomes:

$$P'_E = \frac{1}{1} \frac{\pi^2 EI}{h_o^2}$$

Fig. 4.46 Curve of the
determinant values

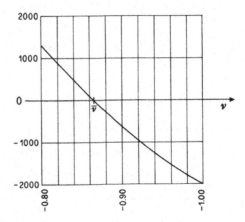

In this way, the ultimate limit state of instability collapse of the frame of Fig. 4.42
has been calculated, in terms of the load p on the structure, in the field of an elastic
behaviour of the material.

Bibliography

Armenakas AE (1988) Classic structural analysis: a modern approach. McGraw-Hill
Au T, Christiano P (1987) Structural analysis. Prentce-Hall
Azar JJ (1972) Matrix structural analysis. Pergamon Press
Bažant ZP, Cedolin L (2010) Stability of structures: elastic, inelastic, fracture and damage theories. World Scientific Press
Beaufait FW (1977) Basic concepts of structural analysis. Prentice-Hall
Chajes A (1974) Principles of structural stability theory. Prentice-Hall
Chajes A (1990) Structural analysis. Prentice-Hall
Chen WF, Lui EM (1987) Structural stability: theory and implementation. Pearson Higher Ed.
Coates RC, Coutie MG, Kong FK (1972) Structural analysis. Nelson
Dym CL (2004) Structural modelling and analysis. Cambridge Univ. Press
Gere JM, Weaver W Jr (1982) Analysis of framed structures. Van Nostrand
Gerstle KH (1974) Basic structural analysis. Prentice-Hall
Ghali A, Neville AN, Cheng YK (1972) Structural analysis. International Textbook CRC
Godbole PN, Sonparote RS, Dhote SU (2014) Matrix methods of structural analysis. PHI Learning
Hall AS (1961) Frame analysis. Wiley
Hibbeler RC (1995) Structural analysis. Prentice-Hall
Ketter RL, Lee GC, Prouwel SP Jr (1979) Structural analysis and design. McGraw-Hill
Mau ST (2012) Introduction to structural analysis: displacement and force methods. CRC Press
Mc-Guire W, Gallagher RH, Zieman RD (1979) Matrix structural analysis. Wiley
Morice FB (1959) Linear structural analysis. The Ronald Press
Norris CH, Wilbur JB, Utku S (1976) Elementary structural analysis. McGraw-Hill
Pilla DR (2018) Elementary structural analysis and design of buildings. CRC Press
Prevost JH, Bagrianski S (2017) An introduction to matrix structural analysis and finite element methods. World Scientific Press
Przemieniecki JS (1986) Theory of matrix structural analysis. McGraw-Hill
Spillers WP (1972) Automated structural analysis: an introduction. Pergmon
Thadani BN (1964) Modern methods in structural mechanics. Asia Publ. House
Timoshenko SP (1956) Strength of materials, parts I and II. Van Nostrand
Timoshenko SP, Gere JM (1961) Theory of elastic stability. McGraw-Hill
Timoshenko SP, Young DH (1965) Theory of structures. McGraw-Hill
Tuma JJ, Munshi RK (1971) Advanced structural analysis. McGraw-Hill
White JW (1978) Advanced structural analysis. Granada Publisher
Zienckiewicz OC (1977) The finite element method. McGraw-Hill

© Springer Nature Switzerland AG 2019
G. Toniolo, *Introduction to Frame Analysis*, Springer Tracts
in Civil Engineering, https://doi.org/10.1007/978-3-030-14664-1

Printed in the United States
By Bookmasters